T0329000

People over Process

People over Process
Leadership for Agility

Michael K. Levine

Routledge
Taylor & Francis Group

A PRODUCTIVITY PRESS BOOK

First edition published in 2020
by Routledge/Productivity Press

52 Vanderbilt Avenue, 11th Floor New York, NY 10017
2 Park Square, Milton Park, Abingdon, Oxon OX14 4RN, UK

© 2020 by Michael K. Levine

Routledge/Productivity Press is an imprint of Taylor & Francis Group, an Informa business

No claim to original U.S. Government works

Printed on acid-free paper

International Standard Book Number-13: 978-0-367-34188-6 (Paperback)

International Standard Book Number-13: 978-0-367-36990-3 (Hardback)

International Standard Book Number-13: 978-0-429-32475-8 (eBook)

Library of Congress Cataloging-in-Publication Data

LoC Data here

**Visit the Taylor & Francis Website at
www.taylorandfrancis.com**

Contents

SECTION 1   Introduction to Facilitative Leadership for Agility

SECTION 2   Three Major Frameworks (Architecture, Plan, Team Structure)

List of Figures

Preface

People over Process. Or more fully, People and Interactions over Process and Tools. It's the first agile value, from the Agile Manifesto that kicked off the agile software revolution almost 20 years ago. And yet, it seems that discussion on agile software is dominated by Process, Scrum, Kanban, SAFE … but what about the people?

This is my third book about lean and agile software. The first (*Tale of Two Systems*) was about the process – why do we do lean and agile approaches, and what are they? The second (*Tale of Two Transformations*) was about organizational transformation – how do we make our organizations do lean and agile development? In both, I spoke to the primacy of the people involved, but focused instead on process and organizational change. This book, the final in this lean and agile trilogy, focuses in on the first principle, the people who do the lean and agile software development work, and the key topic to increase the odds of success: creating a culture of facilitative leadership.

I walked into my first training session on The Facilitative Leader ready to learn, even though I didn't know it. I'd been leading large technology organizations for about 10 years and had just taken on my largest technology and business operations role yet, moving from the 250-person range up to a thousand or so. At the smaller size the skills that had worked well for me – reportedly smarter than your average bear, a good communicator, hard work and honesty, financial and technical knowledge – were enough to generally succeed. My typical pattern had been to figure things out and orchestrate delivery. I was "intellectually arrogant" and thought being smart and driving execution was enough; figuring out what to do and driving its execution was leadership.

When Lucy, my leadership muse, asked us to define the role of the leader, and what we typically wanted to accomplish in meetings, my thought was "how do I get these people to help me figure out what to do and then execute upon that?" A few days later I had a better intellectual understanding of leadership for agility (although I didn't call it that yet), but it took me some time and some experimentation to incorporate the new behavior into my default style (and I'm still trying!).

Lucy's training exposed me to the basic ideas and techniques upon which I will elaborate in the pages ahead, but it was my team members who showed me how useful more effective leadership could be. First, it was a peer in an offsite breakout session, who volunteered to facilitate the session and helped us achieve useful results (which as anyone who attends many of these corporate events knows is somewhat rare). Then some of my team members created the architecture simulation event, a true breakthrough approach to creating and exploring the interaction of solution design with business process. As we trained more people in the ideas and techniques, it was compelling to see the improvements in both business results and team member satisfaction within the organization.

Over several years at two major institutions, and in other forums, I've seen first-hand how people can learn to lead and the accelerating value better leaders provide as the foundational culture of interaction improves. In this book I want to share it with you.

All the best,
Michael K. Levine

About the Author

Michael K. Levine brings together compelling insights on lean and agile software development, creative writing and teaching, and decades of practical successful application as a senior manager in leading financial and software companies.

Michael's career has primarily focused on how to profit through the application of information technology. He was educated at Carleton College and Princeton University. His early career was not directly in technology – first he did international trade negotiation at the US Commerce Department and then joined First Bank System as a corporate banker. In both positions he drifted towards software and in 1990 he joined Norwest Bank and began his full-time technology career. Michael progressed through strategy, project management, CTO, and divisional CIO roles, adopting agile techniques before they were so named. He fully embraced lean and agile concepts as they were popularized – helping build and transform teams, and delivering value time after time.

Michael's engagement with modern lean and agile development began at Wells Fargo, where after succeeding in one area with agile concepts he became responsible for a major portion of an enormous and challenged project that did not use these techniques. He then led technology and process engineering throughout the mortgage crisis, rapidly building teams and frantically deploying technology to help struggling borrowers, investors, and the bank. In 2011 Michael joined US Bank where he initially led a program to deploy a new branch banking system (bringing lean and agile to a highly structured waterfall environment) before returning to the mortgage industry as the technology lead for US Bank Home Mortgage in 2013, and now leading all consumer lending and business banking technology.

Michael shared his insights in two well-received books on lean and agile software development from Productivity Press. The first is *Tale of Two Systems: Lean and Agile Software Development for Business Leaders*, emerging from his frustration with unnecessary and expensive project failures and his desire to help avoid such failures in the future. His second book is *Tale of Two Transformations: Bringing Lean and*

Agile Software Development to Life, which focuses on how to change an organization to use the approaches he illuminated in his first. Michael's trilogy is now completed with *People over Process: Leadership for Agility*, which guides readers in honoring the first Agile principle to create and sustain agility.

Michael lives in St. Paul, Minnesota, with his lovely wife Holly Lindsay, who has been patient as Michael spent hours in the basement writing.

Website: TheTalesofAgility.com

Introduction

This book aims to help participants in agile software development environments learn to become leaders. Because this facilitative approach was not native to me, I had to learn it and in my normal (perhaps over-analytical) approach, to document, structure, and explain it. I've had 15 years of experience learning, teaching, and observing its application in agile, waterfall, and transforming software processes, and seen the interaction of good leaders and extraordinarily well-prepared meetings result in remarkable achievements. Less because I'm a good facilitative leader than because I struggled to learn and apply the techniques, I believe I have distilled the essence of good leadership for lean and agile software and can help you learn it.

Facilitative leaders should be at every level of the organization, from individual contributor to informal team leader to managers of all stripes. That takes a lot of focus and intentionality from senior organizational leaders, who have special obligations in creating successful lean and agile development environments. I dwell significantly on this topic, with a chapter devoted to those obligations, and the leaders of our fictional Pacifica Bank case study demonstrate this kind of leadership in action.

Beyond the principles of facilitative leadership for agility *People over Process* provides tips and demonstrative scenes for some of the more important and common software meetings: architecture simulations, project planning, team configurations, retrospectives, and more. My intent is to bring the principles to life for you and to share proven techniques for the most important leadership events in agile projects.

While this book, like the training I received, focuses on facilitating extraordinarily well-prepared meetings, I want to emphasize that to some extent this is a metaphor for leadership more broadly. The leader's obligation to help their team make rigorous fact-based decisions; to gain broad input and have participants aligned on the outcomes and next steps; and to do so in an efficient way that respects the time of the participants, is as relevant to every-day leadership activity as it is to conducting meetings.

I've written this book following the basic model from my two earlier books on lean and agile software, *A Tale of Two Systems* and *A Tale of Two Transformations*. I mix background and explanation with demonstration, in this case the story of an agile project at Pacifica Bank. The bank and all participants are fictional, although inevitably I draw on experience and archetypes from my many years in the business. The scenario I constructed at Pacifica was done to illustrate the concepts – I needed a project that had a good grasp of business needs but not much of a plan to do the development work. I also needed to create a juxtaposition of a pure process-follower (in this case, the agile coach) and an advocate for facilitative leadership (our star Mary O'Connell). Some of the elements of the scenario and some of the debates are not going to ring totally true, so I ask for patience with anything that seems contrived – the goal is to explain and show leadership, not to tell a great compelling story of a software delivery team and its struggles. If I have created a story that holds your interest while learning about leadership, which is what I've tried to do, it's a bonus, but you might occasionally have to willfully suspend disbelief (I hope not often).

I try to summarize the main ideas, show them in action at Pacifica while commenting on key elements for your attention, and then provide specific tool and technique ideas for you to come back to in the Appendix. In the background sections I illustrate the ideas with stories from my own experience to try to bring the concepts to life a little better.

The focus of the book is on the flow of software from understanding what is needed (we touch only lightly on this) through design, development, testing, and deployment. It is definitely not a book on customer needs, product management, software architecture, testing methods, or software technologies. The focus is on leading the people involved in these activities, especially at scale.

When I explained to my software-engineer son Sam what this book is about, he asked with his usual insightfulness, "Dad, aren't you just prescribing Process for how to lead People?" Sam's question cuts to the heart of what I hope the message is here: I am providing a simple and powerful model of leadership and examples and tips. This is not a cookbook on how to lead. It is a set of principles and examples. Every leader needs to find his/her own way for their team, their organization, and their unique challenges.

I hope the exposition works for you and helps you become a better leader and helps your organization – and all the people of whom it is composed – become better at lean and agile software delivery.

Section 1

Introduction to Facilitative Leadership for Agility

- In this section the leadership model is introduced for both initiative leadership and organizational leadership.
- Our fictional case study, Pacifica Bank, has begun a major technology product development project. The Bank and the team are introduced, and we see the project slightly off track.
- Our archetypical facilitative leader, Mary O'Connell, diagnoses the issues and helps the team get back on track. In so doing the hiring manager, Sai Kapoor, demonstrates some of the organizational leadership principles that have been introduced, while Mary demonstrates facilitative leadership behavior appropriate to the Bank's important project.
- The important role of meetings is explained and techniques to prepare for and conduct extraordinary meetings are elaborated upon. The section ends with Mary conducting a course correction meeting and putting the project onto a better path.

1

Pacifica

"It's Not Agile If There Is No Software!"

Mary O'Connell sat by herself in one of the huddle rooms overlooking the new agile studio. It was late Thursday afternoon on a cool, Southern California, January day. After four busy days of observing and talking with the Pacifica agile team, Mary was processing. She twiddled her pen nervously as she considered what she had seen this week and pondered what to do about it.

Mary had two topics on her mind as she looked over the emptying studio expanse as the sun set behind the coastal hills to the west. The first was formulating conclusions on what was happening in the hip new space below her; the second was deciding what to do about it.

Three months into its conversion to agile, Pacifica Bank's studio team was energized and excited. Guided by Davidson & Guilderson (D&G), a major international consulting and integration firm, Pacifica was attempting to make the leap to digital relevance by junking its cubicles, ending its silos, and turning its view from internal barriers to true customer needs. From what Mary had seen in the last few days, the feelings were real and progress was well underway, but a tweak in direction was needed.

The studio had all the modern elements of an effective team space and was well along the road of a customer-focused agile program. The walls were filled with personas representing Pacifica's target customers including Jenni the mid-market US CFO, Hiro the Japanese export broker, and Sanjeev the Indian treasurer of an importing retailer. A state-of-the-art customer engagement center stood next door, where the team had engaged with persona stand-ins to learn what they thought about Pacifica's and competitors' products and services.

Decisions had been made about tracking and management tools and those tools were now filling up with user stories describing what the team had learned was needed in Pacifica's new mobile apps. Every day the team gathered for the daily standup as the loudly ticking round clock on the back wall ticked past 9:00 AM, right after the catered free breakfast drew them in. Yet, something was wrong, and Mary now saw it clearly. She was encouraged that a few of the participants in the studio and at least one of the managers engaged in the light governance group had a similar sense of unease.

Mary was also encouraged that Pacifica's executive management knew enough to ask for an independent review of their progress. Thus, her presence, recruited for this gig by an ex-teammate now at Davidson. Pacifica's EVP of Commercial Banking, Sai Kapoor, had engaged and trusted Davidson to help him, but he was just skeptical enough to want more than one view. Rather than risk dissonance with D&G, Sai had asked the partner on the account, Melanie Strom, for a referral for a second opinion. Sai had also conferred with Heather Gilliam, the Pacifica CIO, who had agreed on the risk-management approach.

It was through Heather that Mary had been introduced to Sai. Heather had met Mary a few times at local technology events and had been impressed with her work at Mary's previous employer, FinServia. At FinServia, Mary had led a complete transformation from outsourced rigid waterfall development to an in-house team following agile principles, with dramatically improved results. Heather had heard Mary speak at an event about both the need for lean and agile techniques and about how to adopt them effectively. Heather had since hired a few of Mary's ex-team members as their careers matured and liked what she had seen and heard. When Sai asked for a recommendation Mary was the first name that came to mind. (Note: Mary's full story at FinServia is told in *A Tale of Two Transformations: Bringing Lean and Agile Software Development to Life*, Michael K. Levine, Productivity Press, New York, 2012.)

It wasn't often that a senior delivery-oriented executive like Mary was available as a consultant. Fortunately for Pacifica, Mary had been away from the work world for almost two years having her first child and was contemplating looking for a part-time way back in. She jumped at the opportunity to help a local company embracing the values and principles she championed and came on as part-time D&G consultant. She knew she could help, although she was nervous about the new role as a consultant rather than a direct leader of the work.

Mary must have heard the word "agile" 100 times in the last few days. Superficially, the agile studio had all the external characteristics one might expect. But one thing was missing – the software! No matter how much the people enjoyed working together, engaging with customers and thinking through what was needed to truly compete against earlier digital adopters, there was no software flowing and no real plan to get it going. One of Mary's favorite agile principles from the now aged Manifesto (see Appendix 1: Manifesto for Agile Software Development, page 255) was "working software is the primary measure of progress." In the room below, this meant no progress had been made. This wasn't entirely fair, as Mary knew that the study period that should proceed jumping into a development project is critically important, and this period was progressing generally well. Nevertheless, the lack of concrete progress towards software was disturbing.

How could this be? Davidson had plenty of experience helping organizations set up what seemed to be agile processes. Pacifica had made a major commitment to the program. Pacifica's people had done what was asked of them: learning about scrum, studying competitors, listening to customers, compiling user stories into their backlog in their agile management tool. But Mary saw, after just a few days, the missing ingredient: leadership. The team was stuck making the transition from study to code and seemed unable to break through. It seemed to lack either the expertise to apply the scrum methodology to its circumstances in an agile way, or the collective capability to apply its expertise and experience together. Or both.

Mary knew that her obligation was to help – both to get through this modest directional challenge, whereby what had become the standard agile scrum process was not a perfect fit, and to help the team build its own leadership muscle.

| Signposts | Mary O'Connell is a highly experienced development leader, expert in lean and agile software. She has returned part-time after an extended break to have her first child. She has been asked by Pacifica Bank to evaluate their new agile development project, working with their primary consulting firm D&G. No warning signs are yet flashing for Sai Kapoor, the Pacifica business leader, but he wants to be sure Pacifica is doing the best they can. |

Leadership Guides

- When embarking on a new way of doing things, take advantage of expertise to get more than one perspective on the path.
- Avoid creating inherent conflicts among experts on your teams, as Sai has done by including D&G in the selection and management of their second opinion provider.
- Building complex technology solutions is about much more than just agile processes and excellent leadership. Towering technical expertise like Mary is bringing is foundational to success.

Coming Up Next

- We break from Mary's consideration of next steps at Pacifica to explore the nature of leadership needed to apply agile principles effectively. Agile's heritage and principles don't speak directly to leadership, so we will develop a model of agile leadership as a valuable ingredient to success. First, we will consider people and initiative leadership in general. Then we will consider the special case of the organizational leadership for agility.
- Following our backgrounders on agile leadership, we will return to hear Mary's diagnosis of Pacifica's situation.

2

Background

The Facilitative Leader for Agility

THE EMOTIONAL ROOTS OF AGILE

The *Agile Manifesto (c. 2001,* see Appendix 1: Manifesto for Agile Software Development page 255*)* was so named intentionally, as a revolutionary document aimed at restoring the centrality of the software developer. Developers, the Manifesto implicitly declared, are not just resources that can be outsourced. They aren't a cog in a machine fed through contracts and documents. To be successful in software development, the Manifesto declares, you must be developer-centric. Some of the declarations are right to that point.

- We value people and interactions over processes and tools.
- Business people and developers must work together daily throughout the project.
- Build projects around motivated individuals. Give them the environment and support they need, and trust them to get the job done.
- Working software is the primary measure of progress.
- Agile processes promote sustainable development. The sponsors, developers, and users should be able to maintain a constant pace indefinitely.
- The best architectures, requirements, and designs emerge from self-organizing teams.

The values and principles were a scream of frustration against the horrors of Dilbertesque corporations by self-described software anarchists. It was a reaction against corporate pursuit of reliable and low-cost software development typified by infatuation with the Capability Maturity Model, offshore outsourcing, and productivity measurement by counting lines of code and/or function points.

The Manifesto struck an enormous chord of sentiment in the industry (and in me). That resonance has echoed through the last 15 years, and as a result agility in technology has become common if not dominant. Clearly the anarchists had insights that mattered. Many of the ideas work.

However, we are left with a challenge for those of us who actually manage these Dilbertesque organizations. Are we consigned to find great people, put them on a team, and hope for great results from self-managed teams? Is that the extent of our responsibility? Or do we have a higher calling?

Let's consider one other perspective on the agile revolution. It wasn't just that the clueless corporations disrespected creative artists and teams, although the pursuit of turning software developers into automatons was in full mistaken flight. That goal, of prescribing process and measurement and standardization, reflected a fundamental misunderstanding of the current nature of the large-scale software development process itself. Before we address the obligations of organizational leaders for agile endeavors, we need to have a clearer understanding of software processes.

PREDICTIVE VERSUS ADAPTIVE PROCESS CONTROL

In my first book, *Tale of Two Systems*, I explained how this agile rebellion was a reaction to a widespread confusion about the nature of complex software development. There was a strong strain of belief that software development could be turned into a reliable factory, with requirements going in, a series of repeatable and improvable steps being executed, and great software emerging reliably out the end. After all, if an automobile factory could turn out a high-quality car every minute, shouldn't we be able to engineer software development into a reliable manufacturing process? The idea makes great sense but kept running into reality: the larger and more complicated the project, it turned out, the less likely success.

The frustration of large-scale failed software development projects was exacerbated by equally large adoption and implementation issues. Large organizations increasingly were able to buy software packages but found that installing and getting their people to productively use the new technology could be as difficult and dangerous as the software creation process.

It was, and is, the unfortunate and long-resisted truth that the best business analog for large-scale software development is not a manufacturing model; it is a new product development model. Figure 2.1 contrasts key features of manufacturing versus new product development. One is highly predictable, measurable, reliable; the other is by its nature a series of experiments and learning and adaptation.

Because the nature of manufacturing and product development are so different, they must be managed and controlled differently. Manufacturing processes are standardized, measured, and continually improved – in software, the five levels of the CMM are the most prominent attempt to promulgate this approach. In contrast, new product development processes focus on developing great people, constructing an environment where they can effectively collaborate and apply expertise, and

Manufacturing	New Product Development
It is possible to first complete specifications, then build.	Rarely possible to create upfront unchanging and detailed specs.
Near the start, one can reliably estimate effort and cost.	Near the beginning, it is not possible to reliably estimate effort. As empirical data emerge, it becomes increasingly possible to plan and estimate.
It is possible to identify, define, schedule, and order all the detailed activities.	Near the beginning, it is not possible to specify all the needed steps. Adaptive steps driven by build-feedback cycles are required.
Adaptation to unpredictable change is not the norm, and change-rates are relatively low.	Creative adaptation to unpredictable change is the norm. Change rates are high.

FIGURE 2.1
Manufacturing vs. new product development.

establishing events or milestones at which progress can be evaluated and necessary adjustments made.

The Dilbertesque nature of corporate behavior regarding software that so provoked the Manifesto authors was in large part an attempt to put into software projects the kinds of process controls, and establish the reliability, that corporations sought (and often found) in managing other activities such as manufacturing, inventory control, financial management, and regulatory compliance.

THE FACILITATIVE LEADER: A MODEL FOR AGILE SUCCESS

Large-scale software development and implementation is a new kind of human endeavor. Getting a hundred or more people to assemble complex ideas into an invisible machine that enables transactions, controls operations, and makes decisions hasn't been done prior to the last 50 years or so. Humans have cooperated to design and build large complex physical structures for millennia, certainly back to the Pyramids or before. And they have coordinated in large-scale command and control endeavors notably waging war. But the realm of creating pure ideas and turning them into code that invisibly runs, thousands or millions of interlocking instructions, is a brand-new challenge. Purely intellectual collaboration that nevertheless demonstrably works or does not is quite novel.

To be successful in this new endeavor, two leadership elements are crucial:

1. Leading teams of people to deliver valuable solutions, and
2. Enabling teams within larger organizations to succeed.

This chapter covers the first element: team and cultural leadership. The special demands on organizational leaders are covered in the next chapter. Together I am calling this model "Facilitative Leadership for Agility."

The Facilitative Leader for Agility model is a simple triangle, as shown in Figure 2.2. The model reflects the elements needed to enable potentially large groups of highly intelligent, experienced, and diverse people to most successfully deliver outstanding results. An agile development team can, indeed must, have multiple Facilitative Leaders, especially in the critical development, testing, deployment, user experience, and program management functions.

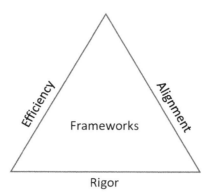

FIGURE 2.2
Facilitative leadership triangle.

The model is a triangle with three sides: rigor, alignment, and efficiency. I'll refer to this as the "RAE" model at times.

I do not claim any right of origin for either the term "Facilitative Leader" or the model itself. The Facilitative Leader term has been used by a variety of people in a variety of contexts (just google it); I am using the term not as a copyrighted model but rather as a description of the kind of leadership needed to achieve technology agility. I would contrast the facilitative style with, for example, "The Process Enforcing Leader," or the "Directive Leader" or the "Risk Managing Leader." These are of course ridiculous contrasting labels, although we've all seen the styles demonstrated in our industry in practice to generally ill effect.

Let us briefly consider the meaning of the three legs of the triangle and the core of it: frameworks.

Rigor. I put rigor first for a reason, in that I believe it to be *primus inter pares* (first among equals) in our model. In dealing with the uncertainties driving agile adaptive process control, there are many crucial decisions to be made. Rigor means clearly defining each decision to be made, gathering and considering facts, thoroughly considering options, and making clear decisions. Without rigor, alignment has to be command driven; while efficiency in pursuit of decisions reached without rigor is just doing the wrong things quickly.

Alignment. Teams must work in a way that gets the best input from all members, and gains understanding and commitment around common goals, schedules, methods, and decisions/directions of all kinds. In this new process of invisibly codifying ideas, perhaps dozens or more team members make many decisions every day that are difficult if not impossible to control. How do we get everyone's head in the game, draw out the best in each team member, and gain strong alignment on the way ahead?

Efficiency. In its simplest form, this means respecting the time of all team members as a valuable commodity not to be wasted. From the humble meeting agenda to a standard format for decision documents to in-person instead of phone meetings, there are proven techniques that can be learned to drive efficiency. Above all it takes leadership that is offended by wasting time to drive the use of these techniques.

Frameworks. Frameworks are the mechanisms through which a leader facilitates a team to accomplish rigor, alignment, and efficiency. Think of a framework as a skeleton on which team members can hang ideas, confident in the flow of that idea towards efficient and rigorous decisions, great software, and ultimate value provision. We will see frameworks at work in the important meetings that set the cadence for this book, on architecture, project planning, and team structures.

In the remainder of this book, we will learn much about frameworks for agility and see Facilitative Leadership for agility in practice at Pacifica Bank, exercised largely through the prism of effective meetings. The Appendix then provides more detail and a summary reference on some useful techniques. Let's now consider the special obligations of the organizational leader, and then return to see some of these techniques in practice at Pacifica Bank.

Signposts	• We do agile software development because we have to – adaptive process control is needed for this kind of work.
	• Despite the emphasis on people in the Agile Manifesto values, it does not speak directly to leadership, naively expecting self-managed teams to be sufficient.
	• Leadership is the critical ingredient enabling successful application of agile principles.
Leadership Guides	• The Facilitative Leader helps their teams with rigor, alignment, and efficiency (RAE).
	• Frameworks are mechanisms used by leaders to help their teams achieve RAE.
Coming Up Next	• We consider the special responsibilities of organizational leaders in achieving agility.
	• Following the excursion into organizational leadership we will return to Pacifica bank and see how our archetypical Facilitative Leader, Mary O'Connell, begins to help the team through their challenging project.

3

Background

The Organizational Leader for Agility

Once we – organizational leaders – accept the need for the kind of approaches that the Agile Manifesto popularized, the question remains: what can and should we do to support, improve, and yes, control our software development and implementation processes? Are we condemned to simply watch and hope, with all initiatives adapting rather than driving visibly to well-understood goals? What is our responsibility?

The classic formulation of agile in the Manifesto has no role for leadership. In fact, it is explicitly anti-leadership, encouraging self-managed teams, reliance on motivated individuals, leaving them alone and trusting them to get the job done. It does appear the sponsoring organization has at least one accountability: to help foster success by giving "them (the team) the environment and support they need."

The most popular technique which claims to implement agile is scrum. While scrum implements many important agile concepts, like the Manifesto itself it has almost no view on leadership. Scrum has the Scrum Master whose responsibility is to conduct specified ceremonies and help the team to overcome barriers. It also has a Product Owner who is to tell the developer what is required to accomplish business objectives. This role is an excellent reflection of agile's roots – developers frustrated by poor connection to business objectives and weak, incomplete, and changing requirements created a magical single-person role that would solve this difficult and complex problem.

Furthermore, neither agile nor scrum contemplates how the agile team should be connected to a larger organization and to external partners who will likely have differing development processes and cadences. Self-managing could work for a small team, but no modern organization will be content for long simply trusting a talented group of individuals alone with responsibility for accomplishing critical organizational objectives. One major failure, or even the looming prospect of failure, and "self-managed" will likely become "highly controlled."

Nowhere does either agile or scrum emphasize the importance of leadership or provide a model of the kind of leadership that would best advance its goals. Even further from the precepts of agile and scrum is a leadership model that would accomplish three critical goals:

- Provide guidance on how the team should balance the dominant paradigm of adaptive process control with the reality that many things in any meaningful project are, indeed, known and can be planned and executed most effectively with prescriptive process control aka "planning."
- Articulate how the agile team itself should be led.
- Connect the agile team to the broader world.

The organizational leader, in addition to driving rigor, alignment, and efficiency in general, has special accountabilities. For Pacifica, we have seen Sai Kapoor in this role. Sai is new to agile organizational leadership but is off to a good start by recognizing the role of expertise and bringing in Mary O'Connell to help.

Let's dive into the responsibilities of Sai and other organizational leaders of agility. Figure 3.1 uses the RAE model to illustrate these responsibilities.

Creating & Sustaining Agility: Responsibilities of Organizational Leaders		
Rigor Making good decisions	**Alignment** Heads in game and moving together	**Efficiency** Respect for people's time
• Right talent, experience, skills, and roles • Team composition • Options considered • Evidence for decisions	• Right involvement • Information available • Input enabled • Value consensus • Someone to decide	• Balance "Agile" and "Planful" management • Frameworks to provide context • Extensive prep for meetings • Tools and techniques

FIGURE 3.1
Responsibilities of organizational leaders.

RIGOR: MAKING GOOD DECISIONS

The organizational leader has responsibility vested in him/her to ensure that each important endeavor, and the overall organizational structure, is configured to ensure rigorous outcomes. We defined rigor as examining issues with multiple viable options, bringing evidence to the discussion, and making fact-based decisions by the right parties.

There are many conditions that breed rigor, but let's focus on three here: ensuring that we have the talent, experience, skills, and roles to enable rigorous work; providing each team with an appropriate set of people, with appropriate skills (including leadership) to make rigorous decisions; and guaranteeing that decisions are made visibly and thoughtfully with viable options and relevant evidence.

Right Talent, Experience, Skills, and Roles

Build Leaders at Every Level

The first accountability of the organizational leader is to recognize the critical role of leadership. In this context, People over Process, in its rightful place as first of the agile values, must be honored and acted upon. This means hiring, training, mentoring, and providing career paths that recognize, encourage, and grow leadership people and roles.

Building leadership starts with hiring. Too many organizations have over-segmented technology management roles, so finding people with the ability to lead cross-functional agile teams is not easy. Organizations just don't create these kinds of people in volume. The fundamental foundation of a technical leader is technical skill and experience – a fact so simple and obvious that it's amazing that it is not universally recognized. Too often organizational leaders, who typically arise through finance or sales or marketing, look for people like them but who seem to have an interest in technology. Technical leaders need deep experience in software development and management – engineering backgrounds, developers, testers, systems analysts. From there project and organizational management can be layered on. By the way, there is a current fascination with "failure as a teacher" – I don't, in general, buy this idea. Look for people with a record of success in successful organizations.

While I'm on the topic of hiring, let me touch briefly on the topic of firing. I don't really like the word or the implications, because as the foundational agile and lean principles state, we must first respect people. Nevertheless, when building and sustaining an organization to execute software in an agile manner, there are inevitably some people who do not fit in a role that must be executed successfully if the broader organization is to succeed. Moving people out of ill-fitting roles is perhaps even more important than hiring, because the implications of not dealing with these situations can affect success on several levels. Typically, a poorly performing team member is no secret; failure of leadership to deal with these issues is taken by the whole team as failure of leadership. As a leader you owe it to the team to deal with personnel problems; no one else can.

Rather than thinking about hiring and firing people separately, let's think of this as a leadership accountability to ensure that the best people we can find and retain are in the roles best suited for their experience, skills, relationships, and aspirations. This might be the single highest obligation of a leader of an organization aspiring to agility.

Growing the leadership skills of the people you have is as important as getting the right people on board. As I explain in the preface, I am a strong believer that leadership can be learned. If I didn't believe that, I wouldn't have written this book! Find training that works – start with senior leaders, and spread leadership expertise throughout your organization.

There is plenty of training available in many companies and in the market for leadership development, meeting facilitation, and tools and processes that support rigor (Appendix 2, page 257, has a nice starting selection). Meeting facilitation training can be particularly valuable, with benefits that run well beyond having better meetings (as important as that is). As we have seen in prior chapters, meetings are a primary mechanism of leadership. The skills that enable excellent meeting facilitation are some of the same skills that lead to excellent leadership in general.

To grow leaders takes leaders. At one point in my career I sponsored excellent leadership training, and we invited business partners in related departments to send some of their team members as well. After a few years of this, as we saw the leadership and meeting culture in one group (call it the Rose group) blossom and business results accelerate. On the other hand, there was no evidence that people in this other group (call it the Crabgrass group) had ever done any training. Meetings led by the

Crabgrass people were as ever, some good some bad, no evident emphasis on clear objectives and meeting paths, no participative exercises, no real change in outcome achievement. It was clear that while the training participants from both Rose and Crabgrass valued the training and learned equally well, the follow-up and expectations from Rose and Crabgrass leaders diverged. Rose leaders had all been through training, knew that they were expected to be facilitative leaders and were evaluated on their leadership skills, and knew that they were expected to grow their team members' leadership skills. Crabgrass leaders didn't take the time to learn facilitative leadership principles themselves, did not practice or expect or support it. The training resulted in sustainable behavior change in Rose, but not in Crabgrass.

What does it take to grow a leader? First, there is learning. Then, there is practice, and feedback.

Recently, a project manager (call him Roger) attended excellent facilitative leadership training. Roger's project was at a stage where an architecture simulation (Chapter 7 Background: Architecture for Agility, page 71) would be helpful, and he undertook to make it happen. He invited his manager Joan to attend as well. Roger felt confident that he could produce this session and worked with another project manager who had experience to help him plan. The meeting produced a few remarkable conclusions, but in general was sub-par as evaluated by Joan. Not unusually, even a sub-par meeting planned with facilitative leadership principles in mind was well-received by the participants. Joan, however, was not pleased, and made sure to give Roger feedback. She started with the positive, as feedback always should, identifying the positive aspects of the meeting plan. Then, without pulling punches, she and Roger talked about some of the weaknesses in preparation and meeting conduct, ending with an agreement that Roger would try again when his project was ready for another simulation and invite Joan again. Joan made herself available to consult on meeting preparation, but Roger didn't feel he needed that and proceeded on his own.

The second architecture simulation was dramatically improved. Roger banned telephone participation, put the agenda/meeting path clearly on the wall and regularly checked in with the team on progress, didn't use any projector at all, got rid of the tables, and had others do most of the work. Dialogue was intense, learning was powerful, and progress astonished many of the meeting participants who had never seen this kind of exercise before. Joan beamed and sent a thank you note to Roger, complimenting him on the excellent leadership and appreciating his willingness to take feedback and improve. Roger was proud of his growth and thanked Joan for her support.

Facilitative leaders often attend meetings and see other leaders in action. This results in impatience with poorly planned meetings and pushback on non-rigorous thinking, failure to get proper input, or running over strongly objecting team members. Good leaders create more good leaders.

Towering Technical Expertise

Several years ago, I was leading a project to develop a branch banking system – the system on the desk of the bankers in the branches to open accounts, take applications, do customer service, sell and market to customers and prospects. It was a good team, following important agile principles and values, using many of the tools and techniques demonstrated in this book. The team members were experienced and knowledgeable. Nevertheless, as we rolled out the system beyond the pilot branches we ran into a serious roadblock: it was slowing down, and despite our best problem-solving efforts we were not making much progress in identifying the root cause and a solution. It was a worrisome day when we had to tell our governance group that we recommended pausing the deployment until we could fix the system.

At this point no amount of excellent leadership, no amount of RAE was going to do much good. We needed the right expertise – what I call Towering Technical Expertise, a phrase I learned from Toyota. Fortunately, one of the members of the governance group, after hearing our explanation of what we'd tried and what the problem looked like, had a suggestion: ask Microsoft for help. Our company had a premier support agreement with Microsoft, and in the past, this had been valuable. In just two days the Microsoft engineer had given us some diagnostic tools to install on our servers, captured and analyzed some log files, pinpointed the problem and solved it. It turned out that our server configuration had settings more appropriate for a workstation, and all we needed to do was change a setting for the number of threads available to a much larger number. We tested it, it worked as advertised, and the deployment continued its successful march across our footprint.

As organizational leaders developing and implementing complex systems, we need to remember that technical expertise matters immensely. Too often organization managers get to their roles through soft skills like relationship management, financial management, people leadership and can undervalue the hard skills that are at least equally important. We need to build the skills specific to our work – software development, infrastructure management, testing, project management, systems analysis, business process engineering – and know when to reach outside our organizations for different or more expertise.

Create New Roles – Chief Engineer

As we've seen, agility requires more intense teaming than we have typically done before, especially in the difficult realm of bringing together technical expertise with business expertise. This new challenge, as you might expect, calls for some new roles. In this section I introduce the concept of the Chief Engineer (CE), the most important new leadership role that must be created and fostered. Think of this role as the technical leader for large initiatives – an entrepreneurial system designer.

In an archetypical waterfall process, the key roles tend to be project manager, product manager, business analyst/user experience manager, deployment manager, developer, and tester. On the technical side the project manager, deployment manager, developer, and tester would each often report into a functional group of similar skills and come together for a project or set of projects. In a strongly waterfall organization that had mistaken models of process control (i.e. following prescriptive instead of adaptive methods), these team members work together by following the prescribed rules. For example, the rules might state that the product manager does the high-level needs statement and business case, followed by the business analyst writing business requirements in a prescribed format, then the developer builds, the tester tests, and the deployment lead deploys.

Once we flip the process control model to adaptive and adopt People over Process, and other agile values and principles, the former guide rails evaporate, at least to some extent. What replaces them? As we've been emphasizing, in general it is leadership. In specific cases, we need to create new roles tasked with technical leadership.

The most important new role I've seen is the Chief Engineer. The conception that I've seen work best is adapted from Lean Product Development, and its practical application to software development processes is illustrated in *Tale of Two Systems*. In brief, think of the Chief Engineer as the senior technical leader on the team to whom all the technology roles are directly or indirectly responsible. The Chief Engineer must be a respected senior member of the technology organization, reporting high enough up in the organization to easily get attention to arising issues of architecture, schedule, resources, or risks, commensurate with the importance of the project. At Toyota, the CE is sometimes spoken of as revered and certainly trusted. He or she must have strong engineering skills as a foundation, layered with project management, communication, business, and leadership skills and experience.

I've not seen this role extant in many software organizations. It is too highly ranked and compensated a role to fit into job evaluation systems that give the most weight to number of people managed and budget dollars accountability, and it fits uneasily between project management and development management. Further, it emphasizes personal leadership capability rather than objective evaluable indicators, soft elements consistent with our basic value of People over Process.

In one organization transforming from rigid prescriptive hand-off-oriented processes to agile co-located teams, the team began operating with just the roles prescribed by naïve agile or simple scrum – product management, scrum master, and development team member. The team floundered in many ways – staying connected to the larger organization, mobilizing help when needed, working effectively together, managing vendor partners. After a few months new roles were added, most importantly a very senior and capable technology manager who was removed from his organizational leadership and assigned to the agile team, plus a project manager, architect/senior engineer, and test manager under his direct control. The addition of this Chief Engineer (in role but not title – there was no existing role or title like this in the organization) made an enormous difference in performance.

Unfortunately for this organization, as they attempted to scale the success of co-located agile teams, the shortage of Chief Engineers became immediately apparent. As senior leaders left organizational management roles non-agile initiatives suffered from the lack of leadership and expertise, and there were no more people like this to assign to new teams. It is exceedingly difficult to hire new Chief Engineers, by their very nature, and waterfall-heavy organizations do not create many of this kind of person.

Short and sweet – it is the obligation of the leaders of institutions pursuing software agility to create Chief Engineers.

Team Composition

For a team to bring rigor to bear, it must have the capability to do so effectively. I'd suggest that some of the required components include:

- People skilled in defining problems and issues;
- Experience in the area at hand to identify options effectively;
- Knowledge of the broader organization/marketplace to identify when help outside the immediate team is called for and how to mobilize it;
- Connection to senior leaders to help judge what decisions should be taken within the team, exposed outwardly for reaction or comment, or simply bucked to decision-makers outside the team; and,
- Effective communication mechanisms.

The balance of skills, presence of roles such as user experience management or test management, and even individual personalities can all make a difference in the success of a team in bringing rigor to bear. The team cannot configure itself effectively without help. Ensuring teams have these capabilities and adjusting over time is a core responsibility of organizational leadership.

A recent example speaks to success in this realm. A large business area heavily dependent on software applications approached development projects by having analysts in the business line "speak for" the business regarding requirements. In practice this meant writing up a bulleted list of requirements and handing them off to development for execution, and then being involved in reviewing designs and testing to make whatever interpretations were needed. As the team adopted agile techniques, this role did not change, just its form – instead of bulleted lists of requirements in documents, the analysts did user stories in a scrum management tool. The results were what one might expect, somewhat mixed; items like regulatory compliance were adequately dealt with, but important areas like user and customer experience were deeply flawed.

Organizational leadership to their credit enthusiastically adopted a new approach for agility, centered on agile studios. The team was reconfigured, co-located, and a new role called User Experience Manager was added. The UX manager spent time with users and customers, arranged focus groups and a/b tests of designs, looked at competitors' offerings, and generally brought a new set of good information to the team. Instead of relying on the overloaded analyst for guidance on user and customer needs, the team now had excellent information from a specialist, including first-hand exposure to input from current and prospective users, some of whom were in the same room as them all the time.

There was no way for a self-managed team to make a change in approach of this magnitude. Shifting to co-location, adding UX team members, re-defining the role of the requirements analysts; all too much for a typical team to do for itself. It took superior organizational leadership to move more rapidly to agility.

Another critical area that must be left to organizational leadership is personnel issues. Teams should be expected to work through conflicts, rearrange roles, and recruit new members as needed. But what if there are performance gaps? How can a team of peers remove a troublesome or inadequate member who formally reports outside its structure, or even give strong messages about required improvements in behavior or results?

I had joined a new company with a big transformative program underway. As I got to know the team and observed its working, status, and plan, it quickly became clear that the project manager just wasn't cutting the mustard. The development manager, the lead of the systems analysts and testers, and individual team members all complained directly about him. They were stymied to fix the issue, however; talking with this project manager had yielded no meaningful improvement. I spent some time with the project manager of concern to get his perspective and take his measure. Ultimately, I decided to terminate his contract (he was a consultant, so it was not as traumatic as if he'd been an employee). The team members stepped up and filled the void, so much so that the event became quietly known as addition by subtraction. Teams cannot usually remove poor fits themselves; that is a responsibility of organizational leadership.

We've seen in the Pacifica story how Sai is executing his responsibility to the team by asking Mary to evaluate its progress, and upon learning of its needs adding Mary to the team to help. In this case the leadership, from Jackson and others, wasn't poor, just not inclusive of the top-notch experience and expertise that Mary could provide. The Pacifica team was stuck, and Sai helped to un-stick it. Good leadership!

Options Considered and Evidence for Decisions

The next obligation I'll explore here is to ensure that decisions are made visibly and thoughtfully. Many leadership actions can contribute to this. Some examples include:

- Refusing to endorse a recommendation if not presented with multiple reasonable options;
- Insisting on evidence for options;
- Guiding team processes.

A recently formed team was working on planning and estimating required interfaces for a new system in preparation for an investment decision. Because the team was new, including several members new to the organization, I asked for an early review of how they intended to do their work. One of the team members had brought an approach from a prior engagement called an Interface Design Document or IDD. This wasn't a bad approach – the document began with a description of the business need, then spoke of the connecting system, and ended with a sequence diagram of the system-to-system interaction along with a high-level description of the technologies involved. In this situation, however, there were several items missing – in fact the IDD would be an excellent outcome of the analysis, but due to broad involvement across the organization around this work I felt we had an obligation to be more rigorous. I asked them to substitute a Decision Document for each interface for the IDD and do the detailed technical design in the IDD later.

What was missing? If we go back to our definition of rigor, it is evident:

- There were no visible options. Many of the business needs could be met in multiple ways, from the users simply toggling between two systems, to connecting to existing systems, to implementing new solutions that better fit with the new technology being implemented. To ensure good decision-making, we had to show our options.
- There was no evidence to support the chosen option. Why should this need be met by this integration? What are the benefits, costs, and risks?
- It was not clear who should be making each decision.

As an organizational leader with significant responsibility for this work, it was my obligation to influence the degree of rigor brought to this work.

I judged that more rigor was needed, guided the team to do the work in an effective way, and monitored progress and success. Leadership for agility in action.

ALIGNMENT

We've explored what alignment means in the overall context of facilitative leadership. Here, we are concerned about how the obligation of ensuring alignment should drive the organizational leader.

Ensure Proper Organizational Context for Decision-Making

As shown in Figure 3.1, three of the identified obligations revolve around ensuring that the agile teams are connected effectively to the broader organizational context: they have the right involvement of people outside of the team, including appropriate communication and governance structures; they have the information they need to make good decisions; and the teams are well-led and culturally enabled so that input from the team and outside of it is gathered effectively.

An example from my distant past is a good illustration. I was Chief Technology Officer (CTO) of a company that both printed Multiple Listing Service (MLS) books and provided online systems for regional MLSs. This dates me – most readers can hardly imagine that in the old days home buyers would visit their realtor to page through thick books of homes for sale! A team within the book publishing group was doing some strategy planning and came up with the brilliant idea to put the books online. They were an empowered team, so they searched for a solution and wound up working to engage an outside vendor to build them an "online book." I heard about this when the Chief Financial Officer brought the contract, in his office for final approval, over to my office, wondering whether the online systems for which we were already a national leader weren't really just the books online? He was right, and we sadly had to tell the team that they were far off-track and the project for which they were all excited wasn't going anywhere at all. This was not necessarily a team failure; it was a failure of their leadership (and probably mine as well) to ensure that they had enough organization context and connection to be on track.

There are both direct team constitution and operating actions, and broader cultural and tool mechanisms through which leaders can ensure the proper connections. Building on the example of decision documents discussed above, the facilitative organizational leader might also insist that the decision documents include identification of the management constituents and how they are involved in the decision. Team governance (which we cover later) is also a crucial framework to ensure proper organizational context without unnecessarily crippling team autonomy and coherence.

Make Decisions Well

The other two rigor items in Figure 3.1 refer to decision-making: valuing consensus and making decisions when needed. It's easy to say that you value consensus but much harder to demonstrate it over time. Spending time with the whole team, not just its leaders, and asking individual team members what they think goes a long way. Perhaps you've heard of the chickens and pigs scrum rule? The punch line is that organizational leaders attending scrum meetings should sit silently and not participate. The team members are the pigs and are truly committed (like the pig is to breakfast, since they supply the bacon with their lives) while the organizational leaders are less so (like the chicken is to breakfast by providing eggs). There is certainly some wisdom in this, to enable the team to work and reinforce their autonomy. On the other hand, by actively participating with the whole team the organizational leader can reinforce the degree to which he/she, representing the organization as a whole, values the team's opinions.

Consensus in the agility context should be thought of as providing all with good opportunities to contribute; transparency of facts, values, and options; and broad agreement with the outcome. It is not the same as unanimity nor is it typically a major-rules vote. Some team members may vehemently disagree with some decisions; that is normal. But if the team members respect the decision-making process (see rigor) and respect the ultimate decision-maker, consensus can be achieved even in the face of strong dissent.

Which points to our final requirement, that someone respected is empowered to make decisions when things get tough. This happens in the face of strong dissent, when values matter more than facts, or when the facts are in dispute. Cross-functional teams can get stuck or start getting into issues that are politically sensitive or for which the right answer might entail unbudgeted costs or risks.

A team had been working on a new system for an important business area for quite some time. The project had been beset by problem after problem, including changes in the business area leadership, poor performance by vendor partners, technical flaws in some critical components, defect-ridden code delivery, turnover in critical roles – you name it, this project suffered through it. At each critical junction the team came up with ameliorative next steps to keep the project moving ahead. Most were tough to swallow, as they involved delays and more spending. But the alternative at each step was worse – take a financial hit from writing off the capitalized costs, further delays in the business line getting needed capabilities, the staff admitting to failure and the feared career consequences.

Another in this series of setbacks arose and the team took its usual path: how could they fix this problem and what would be the implications to time, cost, and risks? The team did a good job within this constrained context – no one on the team had the vision or empowerment to consider the full nuclear option of killing the project and starting over. This is an extreme example of the limits to a self-managed team, regardless of how talented or empowered. Very few of the team members, if any, had enough depth and length of experience to seriously suggest adding this option to the decision set.

When the organizational leader (call her Jody) came to understand that the project was in trouble yet again, she drew on her experience and relationships to other business and technical leaders to raise the idea of terminating the project entirely. Jody also helped structure the investigation of the termination option – how public to make the analysis, how far to go into looking for alternatives, what kind of reassurances, if any, to team members likely to be impacted by termination. She was able to be the leader the team needed at the time, to inject otherwise out-of-scope options and help them re-engage around a somewhat different mission. And ultimately, Jody took responsibility, with team and other input, to make the decision.

Agile methods and training rarely speak to organizational management, but as Jody showed here, the role of making decisions for/with the team is critical to success.

EFFICIENCY

A core part of the organizational leader's accountability is to ensure that teams are operating efficiently. In Figure 3.1 we note that elements of efficiency include promoting the use of frameworks, ensuring extensive meeting preparation, and providing supporting tools and techniques. Less concrete but of supreme importance is helping the organization get its fundamental approach to the work right – the proper deployment of adaptive and prescriptive process management techniques.

Balancing Agile and Planful Management

Almost 20 years ago I started to learn about lean manufacturing. I took a new position that included imaging operations at Wells Fargo Home Mortgage as mortgage volume was growing rapidly. We were struggling with throughput, quality, and cost in what was essentially a manufacturing operation. Several hundred thousand paper mortgage files were delivered in several distinct pieces each month, and we had to separate out some physical documents for immediate delivery to secure storage. Then, we had to run the paper through scanners, while separating out each document type for storage in our electronic files. Finally, we had to store the paper for a period while we corrected any errors in the electronic copies and then ship off to storage or destroy.

Failing to process on time or with quality had consequences – from customers not getting their first payment right, inefficient operations dependent on specific document arrival or retrieval, inability to manage default, and the like.

A colleague wisely hired some manufacturing process engineers and partnered with the Lean Enterprise Institute and marched us down a lean operations path. We taught problem management, standardized work, management by walking around, kaizen events, A3 thinking – many Toyota management routines. I learned that at the heart of lean thinking is the idea of standardized work. Over and over we heard, you can't improve your operation until you standardize it.

In its simplest form, this means defining in detail the full end-to-end work process, implementing that, measuring that, and then improving that. The detailed definition took the form of value stream maps and standard work instructions which we posted at each workstation, and trained supervisors to inspect actual work against the instructions. We measured task time against standards, inventory levels at each station, error rates against standards. Teams were taught that they could not change how they worked without changing the value stream maps and the standardized work instructions; instead they were trained in how to do kaizen events aimed at continual improvement bottom up and top down.

We saw rapid and sustained improvement in results. The more we focused on standard work and continual improvement, the better we got. And despite some misgivings, mostly from our mid-level managers, employee engagement and satisfaction generally improved as they understood they were part of a larger mission, and that their input was sought and respected.

In short, we had a perfect match between an underlying predictable process and a strong predictive process control regime. Define each step, do what is defined, measure it, and continually improve it.

Along with several software development managers, the manager of the imaging operations group reported to me. Let's call her Christine. One of the principles Christine advocated was congruence of managers with the production workers around the lean principles, including standardized work. For a manager, this implied a clean desk, with everything in its optimal place, and no ability to change it without everyone doing so. I just couldn't or wouldn't do it. My work and the work of software managers didn't seem amenable to standardization. I'd seen attempts to do standardization of the development process in highly detailed methodologies; while the methods had lots of good ideas, I'd seen following the cookbook directly fail time after time.

This led me to investigate why standardizing software development was so difficult – was there something different about it, or were we just doing the wrong defined steps? Which brought me to Lean Product Development. At the time I was investigating this, Toyota was unequivocally the world leader in automobile innovation and design, yet it had no standard documented automobile development process. It had very powerful cultural norms and traditions, well-understood roles and organizations which worked together around common integrating events, but participants had much freedom in the way they delivered to those milestones.

This much looser process framework was performing very well, in the same company where every step in the manufacturing process was tightly defined and measured and both incrementally and fundamentally improved regularly.

At root, Toyota's leadership was excelling at this critical obligation of management, which in our software context I've called Balancing Agile and Planful Management. The automobile development process, e.g. designing the Prius, has many unpredictable and new challenges and thus calls for adaptive process control, while manufacturing each Prius has few unknown challenges and thus calls for predictive process control. Get it right, and both kinds of activities thrive; get it wrong, and you get failed waterfall software development projects!

Agile software principles are aimed directly at adaptive process control. Deliver something, test it, adjust. Get people communicating face to face in the most effective way to maximize knowledge, so the current situation can be understood and adjustments made. Partner with vendors instead of commiting to fixed contracts which are difficult to adjust as needed. Hold regular retrospectives. Encourage and enable self-managed teams.

In short, we do agile in software development because we must, not because we want to. If we could standardize it like building a Prius, we would. And when we can standardize, we do!

Here is where it gets difficult for leaders in agile organizations. Many processes are amenable to standardization and prescriptive process management. For example, provisioning new employees with their workstations, or dealing with production system support. Taking requests for new reports in an existing reporting environment is another common example. Others are not, most notably complex new system solution delivery. Most are somewhere in the middle – for example, doing regular releases for a mature system, where "plan the work, work the plan" can work very well.

The trickiest area is in initiatives where adaptive techniques are most appropriate (most software development) – to what extent shall we still do detailed upfront and continual planning versus simply iterating, testing, and adjusting? And how shall highly iterative/adaptive initiatives intersect with more highly planned and controlled processes such as a quarterly enterprise release schedule for which content is planned quarters in advance?

Most software work comprises some mixture of work amenable to predictive process control. Almost all of it comprises much work necessitating adaptive measures. Helping the organization get this right is a primary obligation of agile leadership.

Frameworks

We have and will see how valuable frameworks can be used to drive rigorous thinking, alignment, and efficiency, including great agendas, A3s, release planning timelines, two-by-two matrices. As an organizational leader your challenge is to make a small selection of frameworks cultural.

Organizations have only a limited capacity to absorb and make routine frameworks. While individual teams will certainly create their own frameworks both for routine interaction and for specialized purposes such as the kinds of meetings we will show ahead, organizational leadership has the opportunity to establish a few that enable cross-team work and effective governance.

The Scaled Agile Framework (SAFE) is an emerging agile framework. Unfortunately, in my view SAFE is not so much an agile framework as an enterprise scrum framework, quite a different animal. SAFE is a process framework, not a people framework, in some sense violating the first value of the Agile Manifesto quite directly. That is not to say it has no value; the nature of the Manifesto is that we value more the first item (People) than the second (Process) but still value the second.

True people-first agility demands more focus on the frameworks that enable effective people interaction than on standardized processes. I'd argue that good candidates for organizational framework promotion are extraordinarily well-prepared and conducted meetings (of several types such as those in this book) and A3s for framing initiatives and solving problems.

Meeting Hygiene

It is up to the teams themselves to prepare for and execute extraordinary meetings. But organizational leadership has an important role to play in encouraging, training, and demanding performance.

Just recently I received an invitation to a project status meeting with 265 invitees. No, I'm not making that up! I emailed the project manager and she said that folks can come if they want to; the meeting was how everyone involved in any way were to get their updates if they wanted. I couldn't attend but checked afterwards and found out that 30 people actually dialed in for the meeting.

Here is an obvious case where my responsibility was to end this madness. Not surprisingly this kind of awful meeting management is far from unusual. A facilitative organizational leader, in pursuit of efficiency, has an obligation to curate meeting behavior which can easily get out of hand. Your mission is to ensure that meetings have purposes articulated, good agendas, activities instead of lectures, and outcomes that matter instead of optional status updates with no real purpose. See Chapter 5 Background: Extraordinarily Well-Prepared and Conducted Meetings, page 43.

Tools and Techniques

The remaining efficiency requirement is to provide and institutionalize tools and techniques. A short list to which attention should be devoted would be:

- Space. This area is getting a lot of attention as firms seek agility. Pacifica has chosen to do a dedicated, beautiful mostly open shared space with plenty of room and great wall spaces for collaboration, optimizing the teams' geographic concentration. Certainly, great space can contribute to agility. The debate about how open and what amenities probably has different answers for different people and different teams. My observation so far is that a balance of openness which encourages collaboration and private spaces which everyone needs to some extent is the right answer. I'd suggest that as an organizational leader for agility you rely on a rigorous analysis of options, align with the team, and do so as efficiently as you can.
- Agile management tools such as Jira, TFS, Version One. These can be very helpful, especially for geographically distributed teams. Teams need to be very careful that these tools and surrounding process are not overused and are not allowed to get in the way of efficient team work. A recent example of mine is that an agile team was struggling with how to approach the 750 defects in production. They approached this with their tool and their process: they thought a Kanban approach, dealing with the problems one by one in prioritized order, would be the best way to proceed. They established the list in their agile management tool, broke out a dedicated sub-team, got the product manager to prioritize the defects, and began fixing them. The horrendous fact of the quantity of defects in production (and what that really meant) was not dealt with. So be careful with tools and process.

- Utilize communication tools such as phone and video conferencing, shared document storage, team rooms, whiteboards, and meeting paraphernalia such as stickies and masking tape.
- Use common language, at least to some extent. We don't want to force standardization, that is definitely process over people. But if we can understand what "customer" means (folks who buy our stuff, internal partners, both?) that would be helpful.
- Use available mechanisms to learn and share, such as training, forums, web sites or wikis, consultants and such.
- A well-understood prioritization and financial planning framework is important, just not our focus today.

Much of this must be done while building the organizational framework itself, preceding and independent of the establishment of specific agile teams. It's about preparing the garden and watering and fertilizing; if you create a good environment for agility, success is more likely to grow.

DON'T FORGET EMOTIONS AND BELONGING

We've now dealt with supremely logical topics including ensuring fact-based decision-making. But team members are more than logical pools of knowledge and tasks and judgment; they are emotional, connection-hungry human beings, for better or worse.

I once had a direct report, Phillip, who could be harsh with others in a way that undermined commitment and performance. When I counseled him on this, he was resistant to my request that he recognize the role of emotions at work and explicitly deal with his and others'. I believe Phillip's exact words were to the effect that his peers needed to man up; emotions have no role in the workplace, everyone should just do their jobs. Phillip had some learning to do if he was to become a great leader.

Parallel to this encounter are times when recipients of poor behavior search for how to respond. My favorite example was an incident I endured at mid-career. I was CTO of a software company just back from a team building offsite with my peers. We had voyaged to the Colorado mountains, I think Beaver Creek or Aspen. We did an exercise where we climbed a wall, rigged out in harnesses and helmets, the whole deal. Two of us were relatively younger and fitter, and so I was chosen along with a partner to be the climb helpers. The helpee chosen was a Garland, a slightly older, portly, less fit sales leader from one of our regions. He was blindfolded, and our job was to get him to the top of the wall. The exercise was supposed to teach trust and reliance on others. We succeeded in getting him to the summit and applause was given by all, with commitments to trust and relying on each other just like this when we returned to work.

Just a week later I was flagged by my boss' assistant with some helpful information. Garland had sent an email to my boss complaining about an estimate for custom work given by one of my team members and followed this up with a phone call complaining about me and my team. (Validating the advice to be nice to your boss' admin.) She showed me the email, I followed this up with my team on the matter, and found our estimate to be modest and in-line. As you can imagine, I felt angry and, coming so soon after we'd helped that lazy fat bastard up the wall, betrayed.

I wasn't sure what to do about the situation so, as I am wont to do, talked about it to a trusted advisor. My human resources generalist, who was with us at the climbing exercise, didn't hesitate. I needed, she said, to confront Garland immediately. But how?

My wife had a saying in this situation from her years as a kindergarten teacher, in how to respond to bad behavior. "When you _____, I feel _____, I want _____." As a recipient of wrongdoing, you need to tell the person who did the deed, but you need to talk about your reaction to the deed, not the deed itself or the intentions of the person doing the deed. By making it about you, your feelings, and what you want, it is less threatening and more honest.

Reluctantly I knew what I had to do. I called Garland and said, "When you go complain about a routine quote to my boss instead of coming to me, I feel undermined and betrayed, especially so soon after our commitment to be stronger partners, and I want you to come directly to me on this kind of thing until I prove to you that I am either unwilling or unable to help." Garland was not used to direct open discussion of feelings and it put him off; rather than arguing with me (after all, how I felt was how I felt, what I wanted was what I wanted), he apologized, said he didn't mean to undermine me, and promised to come directly to me on escalations in the future. This wound up strengthening our relationship, not undermining it.

People come with feelings. They don't leave them at home. Dealing directly with them can help agility succeed.

Let's turn briefly to belonging. We all know that people have a strong need to belong, that when they feel committed to a group and vice-versa productive interaction can increase and overall group performance as well. Facilitative leaders actively seek to make this happen while staying within the boundaries of their particular milieu. We've all seen the Japanese corporations that start the day with group exercises or heard of Communist-era self-criticism sessions. No going that far, but an occasional outing, t-shirts, beer pong, or afternoon at the curling rink can truly help. Explicitly focusing on building team identity, without building barriers to other groups, is the trick.

A final idea that I've tried to instill in my teams for decades now, a simple concept I call Generosity of Spirit. This means that we should all assume that others want what is best for our organization and our people, and try to interpret every action through that lens. Only when someone proves to us that their intent is malign can we act on that assumption. Try it!

Signposts	Organizational leadership has important responsibilities in helping agility bloom.
Leadership Guides	• Focus on driving rigor, alignment, and efficiency. • Build the conditions for team success. • Model and demand RAE behavior from your teams. • Don't ignore emotions and the need for belonging.
Coming Up Next	Sai exercises his leadership obligations by listening to Mary's evaluation over morning coffee.

4

Pacifica

Mary's Diagnosis

Mary's initial diagnosis was due at breakfast with Sai Kapoor on the Friday of her first week at Pacifica. Sai wanted to hear it directly from Mary first so that Mary's potential feedback on Pacifica's IT department and D&G wouldn't be watered down out of fear of offending Heather or Melanie. So, offsite it was, out of the way and out of sight.

Mary had arrived at the coffee shop of L'Auberge Del Mar, looking out at the Del Mar beach near her home in the northern San Diego suburbs, a few minutes early so she could be sure of what her message would be. She'd found a table on the back patio facing the beach, in the morning sun so she'd be warm enough on this rather cool spring morning. Mary had made a list of her primary observations and recommendations, starting as she always tried to do with the positive elements. She reviewed her list while she waited for her tea to arrive.

Mary was new to consulting – in fact, this was her first paid engagement. Translating her conclusions on the situation in the agile studio into the best message for her sponsors and then into a proper course of action wasn't coming easily to her. It was difficult for her to balance the degree to which she should simply take the initiative to fix the issues she saw, as she would have done in her prior leadership roles, versus providing guidance so Pacifica could fix the issues themselves, as a great temporary consultant should. She also weighed the degree of her personal involvement and commitment to Pacifica over the coming months versus time with her new family. Balance was the key.

Mary had learned how important this project was to Sai in her first interview with him when they had begun to get to know each other. Sai had been at Pacifica for a few years now. He joined first to lead strategic planning, out of one of the major consulting firms. Sai was as global a manager as Mary had known. He was born into what he described as a trading family in southern India, with deep ties to Britain reaching back to colonial days. Sai attended college in Scotland, worked in sales in the City of London, and then came to California for business school. He sold for an Indian tech/outsourcing firm successfully, but not happily, until he was recruited to become a consultant. Through the consulting practice he had engaged with Pacifica, which wound up hiring him to run strategy and get in line for leadership succession. Sai had now been moved into line leadership for further seasoning and development. If Sai could transform how the commercial bank leveraged technology and engaged with customers, he stood a good chance of eventually leading the whole company. The Agile Studio project aimed at dramatically improving the foreign exchange business was his first roll of the dice, and Sai wanted to load the dice as much as he could.

As the bottom of Mary's first cup of tea emerged, so did Sai, walking out of the dark entrance into the bright light of the café. By this time Mary had her approach solidly in her mind. She liked the people she had met so far at Pacifica and thought the project had many elements of success. She had enjoyed being back in the world of adults and believed she could both directly help the team as a part-time guiding member and teach the team to lead for itself. The first order of business was to get Sai on board with the approach she had in mind, and then work out the details with Melanie and Heather. Then it would be getting the team onto a better path towards execution and stronger self-sufficient leadership. Mary wasn't looking for a job or full-time consulting; this engagement was strictly limited in time and scope.

After typical pleasantries while Sai's coffee was delivered, Sai began the business conversation. "So, Mary, tell me what you think. How is the team doing so far?"

Mary had learned from an early mentor that the best way to start a critique was to start with what is going well. She'd learned that years ago and knew how powerful a technique that was to get people receptive to listening, and she had reminded herself to start this way as she prepared prior to Sai's arrival.

"There are a lot of good things happening in the studio, Sai," she began. "It's admirable that you are open to a new way of working, and the support that Pacifica is providing the team is impressive. The facility is great, Davidson is providing some good upfront guidance, and the team is excited and engaged. It looks like the customer and market research are well-along and the business concept appears sound. I'm a big advocate for this kind of study period before plunging directly into development."

"I hear a 'but' coming," smiled Sai.

"Mmm, yes, there is a bit of an issue I see. The goal of the team is to develop a new software solution for small- to medium-sized companies to better manage their foreign exchange exposure – bringing them a capability that to date has only been available to much larger companies. While I'm not an expert by any means in this market, it seems to me that your folks understand the need well and should be into development by now. But there is no real movement to actually build anything. And while we have a target delivery date of November this year, there is no detailed plan to get us there. Without a detailed plan there is no way to understand our progress and risks."

"It did seem we were moving kind of slowly if we want to get into the market this year," Sai agreed. "My lead on the project is Sara Okada – you've met her, they are calling her the Product Owner. Neither Sara or I know much about software development. We are pretty much dependent on the consultants and the technology department."

Sai gathered his thoughts, and said, "Davidson has us in sprint zero, which they say is getting ready. Are you saying that we are in zero too long?"

"It's not just too long in planning, Sai. It's also what the team is doing, or more precisely, not doing that is the concern. And then there is the deeper implication of why the team is not doing what they should be doing."

"You're confusing me Mary. What do you mean exactly?" asked Sai.

"At this stage in the project we should be homing in on a fleshed-out release plan, showing the schedule of development, testing and deployment. We are not on the way to that schedule."

"Why not?" Sai was puzzled. "We have the D&G consultants and a bunch of developers from Heather's team. Shouldn't they know what to do?"

"You would expect that, yes, Sai, but you have a special case here. Heather's people have been told that the agile studio is a new way of doing things and been asked to trust the D&G agile coaching and learn this new method. Even though a few of them are suspicious that the coach doesn't know what he is doing, they are trying to be good corporate citizens and hold their judgment and cooperate."

"So, the D&G agile coach doesn't know what he is doing?" asked Sai, getting a bit agitated.

Mary didn't want to throw D&G under the bus, so she chose her words carefully. "I wouldn't say that exactly. The D&G coach's name is Jackson Maxim. He is a bright kid but doesn't have much practical experience. He and D&G are coaching a methodology called scrum, which is a popular way to implement many agile values and principles. Unfortunately, our project has some complexity to it that will require some adjustments.

"This is why we are stuck – Jackson's approach isn't a perfect fit, and some of the team's natural leaders see that but aren't saying so. Instead of rigorous dialogue, we have some passive-aggressive behavior. Since the method doesn't fit well, the team is not moving forward, so to keep busy they are over-planning requirements in sprint zero instead of moving forward. Many team members are used to doing all the requirements upfront from their prior waterfall experience so no alarm bells are going off – yet."

Sai was absorbing the information. He remembered the team training aimed at helping them break out of old behaviors. It made sense to him that the team would try hard to comply with the guidance from the coaches.

"That doesn't sound terrible, does it Mary? You can help them make a few adjustments to the method and we can break through and get going?"

"It's a little deeper than just the process definition, Sai. Do you remember the first agile value from your training?"

"Is this a test?"

"Sorry, Sai … it's that in agile we value People and Interactions over Process and Tools. It's not that process and tools are not important; of course, they are. But we need to focus on the people on this team and their interactions if we want to really help it succeed."

"That makes sense Mary. If you put them on a better path they could easily just fall back off, couldn't they?"

"You got it Sai. The core issue is that this is a new team and it doesn't have the right technical process expertise to guide them, or the leadership capacity to mobilize the team's knowledge and experience to guide itself. We need to help them by adding some expertise and enabling them to rigorously think through problems, get everyone involved and aligned, and be more efficient."

Sai had absorbed Mary's diagnosis and trusted it. It rang true to his Spidey sense that something was off. The question now was what to do about it.

"How do we address this, Mary? We have a bunch of agile principles up on the wall that we've been preaching, like 'trust the team to make decisions fast' and 'light governance.' If what you are saying is true, do we have to violate these tenets and step in and re-direct the room? That doesn't seem right," responded a puzzled and disturbed Sai.

"Those principles are great, Sai, but incomplete. You and Heather and Melanie have responsibilities as senior organizational leaders to ensure that the team has the capability of making good decisions. You've made a lot of good steps and the team is off to a great start. But they now need some help. And it isn't just guidance on process; it's learning how to lead in this new agile world."

"Mary, I believe what you are telling me, but I hate the idea of directing the team. Doesn't that violate some sort of agile principle as well?"

"It does, Sai – there is an important principle about the benefits of self-organizing teams. However, remember that agile principles are incomplete – the values and principles don't speak to leadership, either team or larger organizational leadership. For example, there is no agile value or principle that speaks to your obligation as a corporate leader responsible for the costs and benefits of the team in that room."

"True. I never felt right that my job was to put people in that room with some coaching and hope we got a good result at some time in the future. I was feeling guilty about my lack of trust, like I wasn't trusting as I should. But I got over that at least a bit, as you know, since I asked you to come give me a second opinion."

"Entirely appropriate action for you to take as a leader, Sai. I believe the first obligation of an agile organizational leader is to ensure the team is set up to succeed. This includes first and foremost that the people, process, and tools are appropriate to the mission at hand. You have nothing to feel guilty about; rather, you should be proud of yourself for trying to help the team."

Mary could tell by Sai's expression that he was processing. She tried to give Sai an analogy to help.

"Sai, think about another business mission that more obviously requires expertise … let's say building a bridge. If you look at that from the perspective of the bridge's owner, or the firm hired to build it, perhaps that would help. Can you imagine taking a bunch of construction engineers and workers and asking them to work together to design and build a bridge, especially with limited supervision, unless you were absolutely sure they had the right expertise, leadership, and design/build process? Of course not!

"Building a complex business system can be no less demanding than building a physical bridge. In some ways it is even more difficult – the best bridge builders have built many of them, civil and mechanical engineering has been a discipline for centuries, and the techniques, costs, and schedules tend to be highly predictable. For us, there is much more uncertainty. This team has never worked together before, we have a new product we are building with new tools and technologies, and the engineering practices for software solutions are new and changing rapidly. This uncertainty is the primary reason we resort to agile principles. We as agile leaders have an even greater obligation to ensure that the team is prepared to succeed, and to stay close enough to help make adjustments the team cannot do for themselves."

Mary's example helped. Sai wasn't entirely sure about this reassurance but went with it. "OK, Mary, let's say that you and I can agree on how the team's approach needs to be modified to help them succeed. What would you suggest as next steps?"

"First, you need to be sure you understand what needs to be done. In this case, I'd suggest we spend the next hour talking through the turning point at which the team is stuck, so you have a clear picture. Then, you should identify if and how you want to validate the diagnosis. Finally, we need to determine the best way to teach the team and get adoption, and then make the change happen."

"This agile stuff is hard, isn't it Mary?" Sai looked at his phone to check his calendar, sent a text to his admin, and ordered another cup of coffee. "You can help us fix this, Mary? You OK working with us a while?"

"I'd be happy to help, Sai. I'm open to working about half-time for the next several months so long as I can have some flexibility if I need it."

"Excellent. Let's assume that you are going to be helping. Give me the elevator speech, Mary. How do we fix this?"

"First, we need to add some more expertise to the team. I can do much of that, and I can work with Heather to see if we can get some talent from the technology department. Then we need to get the whole team more engaged with taking collective responsibility instead of just following the process that D&G has laid out. As the people and team improve and learn new ways of interacting, we need to ensure that we have a good technology architecture aligned with business processes, a great plan, and a team and governance structure that will support our work. And somehow get the team to do the work for themselves."

"Sounds tricky, Mary. Glad you are here to help. We going to make it?"

"You bet."

Signposts	Sai has put his bid for Pacifica leadership on the line with a new way of leveraging technology and engaging with customers. He doesn't want to take any chances so he's asked Mary to take a look and give him feedback on the program. Mary sees a lot of good so far but warns Sai that the top-level approach has some flaws the team doesn't seem capable of fixing themselves. Sai and Mary set out to help.
Leadership Guides	• The first obligation of a facilitative leader of lean/agile software is to be sure the team's foundational approach is a good fit to the project at hand, and that the team has the knowledge and skills to succeed. • Engaging towering technical expertise to get more than one perspective reduces risks and can help load the dice for success.
Coming Up Next	While Mary and Sai discuss how to help the team, we next present background discussion on how to plan for and conduct an effective meeting. Because surely a big team meeting is coming up!

5

Background

Extraordinarily Well-Prepared and Conducted Meetings

If someone asked you to name the most important single process in a large organization, especially an organization deeply dependent on software, what would you say? Budgeting and forecasting? Sales? Product planning? Hiring and people management?

All of these are important, of course, but I would contend that the most important single process, critical to all of these areas, is the much reviled, hated, boring, confusing, and absolutely critical-to-everything meeting.

Think about the human endeavor of large-scale software development and implementation. It seems to me that this is a largely new type of human cooperation. Think about it:

- Typically, a hundred or more people, with a variety of skills, experiences, and points of view, ranging from gregarious sales people to highly introverted database administrators and everything in between.
- These people need to align around a vision of what they are building and why, and turn thousands of ideas into concrete code that either does what was planned, or does not. There is an objective, tangible, testable outcome of their intellectual collaboration.
- These hundred plus people make thousands of independent decisions, many of which materially affect the probability of success. It is impossible to monitor and control the myriad decisions that team members make every day.
- At the end of the construction period, thousands or more people need to be enabled to productively use the software, and the team needs to monitor use and adjust for the inevitable large or small course corrections needed.

What in human history has been like this? Building physical items like the Coliseum or the space shuttle? Doctrinal collaboration in the Vatican? Governing via the Roman Senate or the Politburo? Each probably has some similarities. But I would contend that never before has this kind of collaborative work been so widespread, so measurable in outcome, and so in demand of improved methods of leadership. That method is the topic of this book. And meetings are a primary vehicle for leadership execution.

What if you want to learn about leadership but work on a small team, and meetings aren't a big deal? Don't skip this chapter! Think of the meeting as a metaphor for leading people in general – thinking about objectives of interactions and how to ensure rigor, efficiency, and alignment is valuable with large groups or 1–1 interactions with a peer or your boss.

LEADERSHIP TIP

Getting good at arranging and conducting meetings is at the heart of good leadership.

This chapter will present the broad framework for conducting an effective meeting. Every meeting must further our primary leadership responsibilities of rigor, alignment, and efficiency. While leadership furthering RAE should be exercised and demonstrated in all leadership activities, highly planned and well-executed meetings are often fulcrums of enormous leverage. If we learn to have great meetings, we learn to lead.

There is nothing terribly new or revolutionary here. My intent is to give you a framework to help you organize your meetings, and then show you these techniques in action in the context of Mary's engagement with the Pacifica agile team. In subsequent chapters we will dig deeper into some of the key topics and watch Mary and others conduct meetings demonstrating techniques most valuable to lean and agile software development.

PREPARING FOR A MEETING

Preparation for an important meeting can take more time than the meeting itself. Every meeting should have some preparation in advance, ranging from the facilitator putting together an agenda and including it in the emailed invitation to extensive collaboration to set the objective, prepare activities and props, and ensure the meeting begins with proper expectations and alignment. This chapter, and this book, is focused on leading complex human interactions so we won't dwell on meetings simple enough for an emailed agenda to suffice. Without belaboring the obvious, the key elements of preparation are presented.

LEADERSHIP TIP

By failing to prepare, you are preparing to fail (Ben Franklin). In other words, Agile is not an excuse not to plan!

- **Set a simple and achievable objective.** This cannot be over-emphasized. It must be clear why the meeting is being convened and what the objective is. Sometimes the meeting objective is easy and non-controversial to set. Sometimes it requires some dialogue. Many times, the meeting convener(s) need to check with others on the objectives and agenda to be sure there is enough alignment to proceed with the imagined meeting at this time.
- **Lay out a path to achieve the objective.**
  - The agenda. Ideally this should be distributed in advance, but that depends on the importance, length, and purpose of the meeting. A good agenda is a string of the right activities in the right order. There are many agenda formats. Rather than clutter up this overview chapter, I've put a full description of a favorite format in the Tools Appendix (page 257).
  - Activities. How many terrible meetings have you been to in which the facilitator stands in front of a flipchart and scribbles down agreements and conclusions? Was the team really being rigorous, gaining alignment, and being efficient? I doubt it. We learn by doing, communicate best with each other directly and not through a leader, and align by creating together. The best meetings have activities designed to get the participants to do the work themselves. Activities should be visual, active, and encourage or require interaction. I typically avoid any use of a projector in an important meeting, except perhaps to do a system demo as part of laying a groundwork of shared understanding on which the remainder of the meeting depends. The difference between a group of people trying to decipher and comment on a projected interaction diagram and building an architecture together on a wall is dramatic; so is the

contrast between allocating user stories to sprints squinting at a projected Version One versus arguing heatedly over which column on the wall a blue task square will be placed.

- **Roles and responsibilities.** You walk into an all-day team meeting with your boss and peers. The agenda is a list of topics to talk about, each with a name beside it. Your boss starts the meeting off and calls for the first presenter, say the VP of Finance, who goes over current financials and the challenges ahead. What are you supposed to do about this? Listen? Ask questions so you understand? Volunteer to help on something? Take a contrary view to test conclusions? If your boss comments on the plan, is that a direction or a suggestion or a conversation element?

  As a meeting planner, you have a responsibility to the participants to help them understand their own and their fellow attendees' role. What is the facilitators' role – to conduct the meeting, to challenge thinking, to contribute substance? How about your boss – is he being the boss or trying to be an equal participant? Who is going to document and communicate results? Lay it out.

- **The physical setting.** One of the most powerful Agile Manifesto principles is, "The most efficient and effective method of conveying information to and within a development team is face-to-face conversation." Leaders can implement and amplify this principle in meetings through how we physically arrange the room to encourage interaction. For the scrum ceremony, teams typically do stand ups with no intervening tables or chairs so that the discussion is quick and team members can move to the front and interact with the scrum board easily. Design your room arrangement to support your meeting path. Hint: a large U-shaped table arrangement is rarely the answer.

- **The paraphernalia.** The activities often demand props: index cards and stickies and tape, markers and dots and scissors. A sign of a good meeting culture is a good supply of meeting tool boxes full of these supplies. Large plastic fishing tackle boxes work well.

- **Ensure alignment on the way in.** You never want to start a meeting and get immediate cold shoulders and crossed arms. Much better that key attendees feel part of it and are interested in participating. You have to judge how much pre-meeting discussion, and with whom, is needed. For some key meetings we will have several phone calls, desk walk-bys, or even in-person individual and/or group preparatory meetings to plan out the meeting objective and path. For others farther along the alignment path, little preparatory cajoling is needed. The old joke about having a meeting to talk about having a meeting is no joke.

CONDUCTING A MEETING

Now that you have a great path laid out, start down it. But be prepared to make changes along the way.

- **Make the path visible and start down it.**
  - ○ I'm now showing my years a bit by saying this but print out the full detailed agenda and hand it out. Electronic devices interfere with human interaction, so don't rely on them.
  - ○ Put the meeting path on the wall, in big visible print. Using a flip chart or two and colored markers works well. I think of this like a path snaking through the woods, getting the team from the beginning of the meeting to the conclusion at the end. Kind of like Candyland, for those of you who have that board indelibly marked on your minds. We will see Mary and the team use a simplified agenda on the wall and check marks to track progress. At the end of each stage confirm progress with the team and look ahead to the end of the meeting. Are you on track to meet your objectives? Check for agreement at each transition because you might need to call an audible.

- **Control the dialogue.**
  - ○ Rules and agreements. Early in the meeting establish rules of the meeting or agreements. Less is more here. If you resort to tried (and often failed) rules like "No phones" or "One conversation" and can't/ won't/don't hold the room to that, you've immediately undermined your leadership role and the value of the agreements. Don't do obvious rules – if you can't argue the opposite, don't make the rule. We'll see some good examples in meetings to come at Pacifica.
  - ○ Activities. Activities are the building blocks of a sophisticated meeting path. Activities should be used to identify and evaluate options, plan schedules, understand and adjust technical architectures, retrospect, determine a minimum viable product, and many other purposes. We will examine some activities in depth in future chapters and see Mary and the team conduct several for their project. Always kick off an

activity by explaining its objective, explaining roles, and doing a tool advertisement. More to come on these.

o Parking lots. It is easy to get diverted – going too deep down a rathole you can't get out of with the people and agenda present. Make a visible parking lot and when a topic is getting out of hand, write out a sticky and plant it there. Our rigor principle might compel us to do a little more than just park the topic.

o Dealing with disruptions. One obligation of a leader is to provide team members with an efficient process (we'll say this a few times!). We all no doubt have story after story of meeting participants holding loud sidebar conversations, or absorbed in their iPhone, or disparaging others personally. It is best to think of maintaining order as a progressive discipline. You can start with just walking over to the offender or making a tsk tsk face at them. Escalate by pulling the offender aside at a break and ask for help. If you need to make an example for the whole room, kindly remind the offender to behave more appropriately. Repeat hardcore offenders can be asked privately if they would rather be elsewhere, and as a last resort be asked to leave.

o Adjusting. You've made a good plan, got off to a good start, but now things are not going as planned. Perhaps the first activity took too long, the second activity just didn't work as you'd hoped, or the participants wanted to go in another direction. Here your maturity as a leader gets tested, and you can be sure it will happen often. Go back to the basics: is the objective of the meeting still valid? What revised path do you propose now, and will the team agree on the revised path? Re-establish the new path and start down it visibly.

CONCLUDING A MEETING

Think of wrapping up a meeting as transitioning to executing on the alignment you have worked so hard to achieve.

- **Checking for alignment.** First be sure that you have achieved alignment. Summarize the conclusions, using the visuals you have created. Your alignment check can take a variety of forms; we'll see a few in the chapters to come including fist to five and my favorite, the thermometer.

- **Agree on communication of results.** If your meeting has been successful, you have agreed on something important. You now need to decide on what should be communicated, to whom and by whom. This can be as simple as a brief conversation and as structured as an activity with stickies and masking tape.
- **Set immediate next steps.** Lay out the immediate responsibilities for action.

Enough background on meetings. Very important, but you can always come back and review. Let's return to Pacifica and see Mary conduct her first major meeting, seeking to help the team get on a better path forward and demonstrate the kind of leadership every agile team needs.

Signposts	While Mary and Sai have been planning how to help the Pacifica team get on a better track for their program, we have presented a framework for how to plan and conduct a good meeting. We recognize that meetings are perhaps the most important process in many businesses, particularly in the new environment of large-scale software development.
Leadership Guides	• A key part of leadership is preparing for and conducting meetings. The cliché common complaint that we have too many meetings is mainly false. The true problem is that we have too many BAD meetings. • We need to ensure that meetings are rigorous, create alignment, and don't waste people's time. Setting clear objectives, planning engaging activities, and regularly checking for conflict and agreement are some basic techniques. • Most meetings should end by looking ahead. How will the results be communicated, and what are the next steps?
Coming Up Next	Mary conducts the first major meeting with the Pacifica team, aimed at getting their fundamental agile process adjusted to their unique circumstances and beginning to demonstrate better leadership.

6

Pacifica

The Course Correction Meeting

A week after Mary and Sai's conversation at L'Auberge, Mary was ready to assemble the studio team and try to guide an important change of course. She also wanted to demonstrate how to get the team engaged to manage their own destiny more effectively by modelling leadership behavior.

In the week of preparation Mary had refined and tested the meeting objective with some of the key opinion leaders in the group and laid out a path for the meeting that she hoped would work well. She had then reviewed the meeting agenda and path with several of the participants, making some adjustments and improvements. She had held individual meetings with Melanie Strom, the D&G partner, and with Heather Gilliam, the CIO. Finally, she had circled to review the input and the meeting plan with Sai, explained his role and gained his commitment. The ducks were in the proverbial row. Time to march.

Ten o'clock came quickly and found Mary, Sai, and a dozen of the core team members gathered around a large whiteboard in a corner of the studio. An uncomfortable quiet possessed the group; it wasn't every day that Sai came to an actual working session as opposed to a scripted demo. Additionally, Mary's observation and questions and preparation meetings had created some unease on the team. Mary had worked with the Product Owner, the Scrum Master, the Architect, the Developer, the Tester, and User Experience Manager, and the Agile Coach (they thought of themselves in these roles as capitalized) to identify the best audience for this meeting, which Mary called Exploring the Way Ahead.

Attendee	Project Role	Primary Role in this Meeting
Damien Lopez	Test Manager	Begin developing the Master Test Plan; advocate for test-driven development
David Phillips	Architect	Share state of architecture; advocate for appropriate steps and deliverables for effective tech arch
Gary Sunderland	Development Lead	Ensure proper flow of information from requirements to code, and associated people and organizational needs
Jackson Maxim	Agile Coach – D&G	Provide guidance on methods based on agile principles and techniques
Judy Grossman	User Experience Manager	Ensure that team effectively uses information already developed about user needs, and stays connected as we move ahead
Kayla Robinson	None	Provide perspective based on her experience and knowledge; consider options for future role with this team
Mary O'Connell	Expert Consultant	Facilitator and topical expert. Help team align on current state and path ahead.
Pam Eldridge	Scrum Master – D&G	Help team adjust future work with outcome of this meeting; document outcome (may ask for help)
Sai Kapoor	Executive Sponsor	Connect team to business outcomes and constraints. Make decisions if needed. Guide D&G role/services.
Sara Okada	Product Owner	Connect business goals to the project plan; advocate for profitability and revenue growth
Vivian Lockerbee	Developer	Focus on the software developers. How do they get the information they need to build the right code, and the support to design and test collaboratively and efficiently?

FIGURE 6.1
"The way ahead" meeting participants.

Mary looked around the studio corner as the participants assembled, and then down to the attendee list attached to the written agenda that sat in front of each attendee (see agenda snippet in our samples Appendix on page 257) and reflected on the cast. Mary had made sure she had enough interaction with attendees so she could document the role the person wanted to play, or the role that Mary was asking them to play in the meeting.

Mary saw that Gary Sunderland (Developer) was chatting with Kayla Robinson, the newest member of the team. Mary's conversations with Gary and with Vivian Lockerbee, a bright young developer on Gary's team, had been revealing. Mary found that Gary did not see a clear path forward into development; he was honestly puzzled and holding his judgment. He liked the direct access to customers and business partners in the studio and was excited to be finally getting good requirements. He was not, however, at all sure how he was supposed to turn the requirements into designs and a project plan if he was banned from doing what he'd done in the past. Vivian (just walking in with her coffee) was on the other hand far from sanguine about the situation to date. She was impatient with all the agile ceremony and Mary was quite sure, though Vivian was reticent to come out and say so, that Vivian was contemptuous of Jackson Maxim's (Agile Coach) supposed expertise.

Kayla's presence was the result of Mary's discussion with Heather. Mary had conveyed her belief that the team was missing technology leaders with experience of agile techniques and had asked if Heather had anyone on board who could help the team take advantage of the agile approach that Jackson was teaching. Mary knew there was much good in what D&G was teaching but the ideal and somewhat simple scrum being taught needed to be enriched and adjusted to the specifics of the Pacifica project. Kayla had recently joined Pacifica from the Air Force, where she had been involved in developing and purchasing software for weapons control. Kayla had been recruited proactively by Human Resources and had just become available. Mary liked Kayla's background in the highly demanding quality world of the Air Force, and her years of experience mixing in agile principles and techniques without sacrificing control and program discipline.

As Vivian joined Gary and Kayla, Mary looked across from them at Damien Lopez (Test Manager) who had arrived early but immediately had fallen into deep commune with his laptop. The Test Manager role is often one of the most important leadership positions, Mary knew, and she was concerned whether Damien would be able to or choose to step up and lead. Damien's perspective on the project status was like Gary's. Like Gary, he understood the basic ideas being taught, but couldn't fully see how it would be realized. Mary would ideally have hoped that by now Damien's perspective would be more like Vivian's, more strongly skeptical that the approach being taught would be effective. Damien liked the early articulation of acceptance criteria as part of the requirements process but didn't see how that would turn into tests in the absence of formal test plans and test cases and test data preparation. How was he supposed to get testing done in the same two-week sprints that requirements and design and coding for each user story was completed? It didn't make sense to him, but he was being patient and not speaking up yet.

Mary's thoughts naturally focused on the development team since that is where the actual work of turning ideas into code happened. In this meeting, Mary hoped the developers and testers would engage strongly and take some ownership of the way ahead. Much of her planning was how to make this happen without putting the enthusiastic agile contingent on the defensive, especially Jackson Maxim, the D&G Agile Coach and engagement manager.

Jackson was in a very different place than the reserved and practical technical group. He was enthusiastic and excited about the project, proud of what had been accomplished so far, and had a clear and confident vision for what lay ahead. He had little conception that the technologists were not fully on board.

Mary liked Jackson and admired his optimism and idealistic dedication to agile. Mary found him smart, a good communicator, and boundlessly energetic. He was young, just 30 years old, but already a senior consultant at D&G, and had been given the Pacifica project as his first engagement leadership opportunity. Jackson had an academic technical background from university in Britain but immediately migrated away from technical to people issues. He'd joined D&G a year out of graduate school and succeeded immediately as a junior engagement manager. In his last project, a major effort building web sites for a European auto manufacturer, he'd fallen in love with agile/scrum, got certified first as a scrum master and then as an agile coach. Unfortunately, he didn't have sufficient technical expertise or experience to fully guide this team, and his perception of his role as change agent overwhelmed his need to listen to the technical experts stuck in the old way of doing things. This diagnosis had been the primary topic of Mary's conversation with the D&G partner, Melanie Strom, who was open to Mary's observations and supportive of Mary joining to help Jackson learn and grow.

In the final moment before bringing the meeting to order, Mary smiled at Jackson and noticed that he was sitting with the team members most strongly aligned with him. Jackson was well-liked by Sai's senior leader on the team, Sara Okada (Product Owner), who felt very fortunate that Pacifica was finally devoting serious resources to product development. Sara appreciated D&G's help in getting actual clients involved in the product development process, and in doing intense market research. Jackson also had an acolyte in his fellow D&G consultant, Pam Eldrige (Scrum Master), and in Judy Grossman (User Experience Manager). The respect the business-oriented team members had for Jackson, and the work done to date, was solid reason for optimism.

Mary took one final look around the room to be sure all was ready. The team was seated in a semi-circle around the whiteboard painted wall. She had posted the agenda (see Figure 6.2) on one side of the open wall space, and an abbreviated form of the Agile Manifesto values and principles on the other. She looked down at her phone and saw that it was time to begin, so she tapped Sai, sitting next to her, to begin.

Sai kicked the meeting off. "Let's get this show on the road," Sai boisterously started. "As you know, I asked Mary to spend some time with us to observe how we are doing and give us some feedback. By now you've all met her, and I'm sure have been as impressed as I am with her experience and wisdom. She's been successfully doing agile technology development since agile was first discovered, right Mary?

"Mary and I have had a chance to review her initial thoughts on the progress of our project, and it's fair to say that we both like our start and progress so far. We both think we are at a good point as we wrap up our customer and market research to step back and be sure we have a path forward with which we all agree. Melanie, Heather, and I have asked Mary to facilitate our discussion today, and also to stay with us for the next several months as a part-time consultant, observer, and teacher."

"So glad to have you on board, Mary," Sai smiled at Mary and the team. "It's all yours."

Facilitating this meeting, her first as a consultant, had forced Mary to think carefully on how to set it up and how to position herself. She had spent her career building and then leading large-scale software products. She was a hard driver, more comfortable focused on the objective than the people or the process. As she had grown in responsibility, she had been mentored by Neville Roberts, 10 years her senior, on how to be a more effective leader. One of her regular reminders from Neville was on how to do a meeting, which included defining roles and responsibilities, objectives, and the path or agenda, and getting alignment upfront. Mary's naturally preferred mode was to just tell people what to do since, after all, she knew best most of the time. But that wasn't the best way for a team to succeed. Mary was self-aware and reminded herself before each meeting for which she was responsible to take a moment and prepare, and be the facilitative leader that Neville had labored for. She began.

"Good morning, team. Thanks for allowing me to join you this week, and for each of you sharing your insights, excitement, and some concerns about the way ahead. I'm also excited to join your team and am grateful for the welcome you've all shown me.

"Today we are exploring the way ahead. If you would turn to the left on the board, I've posted the plan for today, which was developed in collaboration with several of this team. We will start with confirming our objective for today, which Sai has introduced. Then we'll do a little meeting hygiene and get clear on roles and rules, just to help our interaction today go more smoothly. We'll really get going by having Jackson walk us through the Current State and the path ahead he proposes, as our agile coach."

MEETING TIP

Write the major elements of the meeting objective and the agenda up on the wall so participants have a visual shared guide. The initial wall agenda should match the written agenda, just not so much detail. The first step in the effective meeting is to confirm or modify the objective and the agenda. The wall agenda is easy to modify, while the written paper agenda remains as a reminder of the original meeting plan.

<u>Agenda: The Way Ahead</u>

o Confirm Our Objective
o Meeting Roles & Rules
o Review Current State & Path – Jackson
o Check Understanding of Progression to Tested Code – Mary
o Agree on Next Steps – Timeline
o Wrap Up / Next Steps / Communication

FIGURE 6.2
Wall agenda for "exploring the way ahead" meeting.

"After our discussion with Jackson, I have an exercise that I will facilitate to check on our understanding of the progression from today's state, in which we have a lot of user stories and acceptance criteria in Jira, through to tested code ready to deploy. I'm expecting that can help flesh out any fuzzy or unknown areas and move us more quickly to a solid plan, if there are any gaps after we complete Jackson's explanations. We will then lay out a timeline and agree on next steps and how we communicate our conclusions to our broader team.

"We don't need to have all the details of each of these activities right now. I promise that before each activity it will be explained to your satisfaction. Right now, I'd just like to check for broad agreement that the agenda looks acceptable. If so, we'll get right to our goal for the day, and if needed we can always adjust the agenda as we go.

No objections were raised so Mary plowed ahead. She knew it would be a tough audience early on due to the topic and some built-up unexpressed opinions. The team was used to giving Jackson wide latitude to drive its activities and paths as they learned agile. Mary needed to raise latent leadership capabilities from the unusually passive group.

"Our goal today is to review where we are in the project and see if we can agree on the best path forward. I've observed that the project is at a turning point towards actual development and believe from my experience that the team could use some time together to explore the path forward. I've talked to most of you in the past week, and at least some of you are eager to have this conversation."

"So, comments on the objective of the meeting." With that Mary turned to a flip chart poster she had prepared prior to the meeting.

Gary, Vivian, and Damien looked up at the chart, nodded and murmured assent. Mary looked to the folks from whom she expected some resistance and found it in Jackson's look of consternation. But no one spoke up.

"Uncomfortable silence, huh? OK, perhaps it will help to skip ahead a little in the agenda and talk about roles. I'll start with me. My role here today is twofold – one, to be an expert on agile software techniques, and two, to help you to collaborate on exploring the way ahead. It can be a little complicated to both facilitate your involvement and provide experience-based advice, so I'll try to keep the two roles as separate as I can.

"It was in my role as an expert and a consultant to Sai, Melanie, and Heather that I observed the need for this meeting. But right now, I'm in my role as facilitator, where my job is to help this team have a rigorous, efficient dialogue, aimed at getting us aligned around the best path forward. I'm sorry that I will be bouncing roles back and forth today but I'll be as clear as I can. Right now, I'm facilitating, and that means you are supposed to participate! Do you agree that we need to review the state of the project and agree on a way forward, or not?"

Mary stared right at Jackson.

Mary had spent several hours with Jackson over the last week preparing for this meeting. She knew he was uneasy about it and saw her involvement as a rebuke or threat to his role as the Agile Coach. Mary desperately wanted Jackson to feel at ease and accept her help and more involvement from the team but believed that it would be a difficult process to get to that point. Until then, Jackson's discomfort could not be avoided, and it would be best for all to get it out on the table.

"I'm honestly not understanding the issue, Mary," Jackson finally volunteered.

"Seems to me we are doing really well. We've got a great backlog of user stories and have started to allocate them to our sprints in the release plan. Once we do that, we'll start sprint one and then we should be onto our cadence to delivery."

"If indeed you are right this can be a short meeting. Is everyone as clear and confident as Jackson?"

A slight pause, and then the normally reticent Damien Lopez (Test Manager) finally added a voluntary voice to the mix.

"Mary, I think the objective is right on target. We need to talk this through. Some of us aren't as clear about what's next as Jackson is. For example, I don't understand how we can have two-week sprints. We've talked about how we will take each user story and finish them in a sprint and debated what finished means. While we have agreed on a definition, that something is finished when it's coded, tested, and largely debugged, I don't know how we can possibly do that for each user story in two weeks. We should talk that over."

Sai pitched in to help Mary at this point.

"Thank you, Damien, for speaking up. I know it's not always easy to have these conversations, especially in a group with me here, so I appreciate that. Even if the rest of you have this all figured out, which it sounds like you may not, I'd like to hear it for myself. How do we go from where we are to where we need to be? I haven't done many technology projects in my career and certainly none in the new agile way. Mary, how do you propose to do this dialogue?"

Sometimes it helps to have and use positional power, thought Mary thankfully.

"Here is what I'd propose we do. We've agreed on the meeting objective, a great start. I always like to do a level set on our roles and meeting rules, if we have any, so we'll do that.

"Then, we will do our review. Jackson is prepared to walk us through our status and the scrum plan he is proposing.

"Throughout, let's keep in mind the agile values and principles that we've learned and are trying to execute on. I've put some of them up on the wall for reference, and we will come back to them as we need to.

"I've already discussed my somewhat confusing dual role, which in this instance can't be helped. Let's go down the list in front of you all and have each of you briefly explain your role here. Feel free to add to what I've captured in our preparation."

Mary had Sai start. He emphasized his connection to the business outcomes, his ability to provide resources including D&G as needed, and his availability to make decisions when needed. As the role explanations proceeded, for the most part the descriptions in the agenda (see Figure 6.1) proved accurate, no real surprises. The only person new to the team was Kayla Robinson. She was welcomed to the team and said she was there to learn and help if she could.

Mary was pleased with this beginning. Having each person be explicit about their role on the team was a good clarification as they began to look ahead together. The warm up complete, Mary moved the team on to the meat of their conversation.

"Jackson, could you explain where we are on the software development life-cycle. Perhaps start with understanding requirements, which seem to be well along, and then what we think are our next steps toward code complete?"

As Jackson started talking, Mary went up to the agenda on the wall, put a check by Confirm Our Objective and Meeting Roles and Rules and pointed to Review Current State and Path. She made eye contact with team members to be sure they understood the path and the step on which they

LEADERSHIP TIP

Be sure the meeting participants at all times understand the meeting path, and where they are on that path. You can do this visually by using the wall agenda and by verbal and non-verbal clues.

had landed. Satisfied with the non-verbal touch, she let the conversation continue undisturbed.

Agenda: The Way Ahead

- ✓ Confirm Our Objective
- ✓ Meeting Roles & Rules
- ○ Review Current State & Path - Jackson
- ○ Check Understanding of Progression to Tested Code -Mary
- ○ Agree on Next Steps – Timeline
- ○ Wrap Up / Next Steps / Communication

FIGURE 6.3
Mid-meeting "exploring the way ahead" wall agenda.

"Sure," Jackson nodded.

We are now in sprint zero, the planning period before we get going on execution. We have been studying customer needs by doing interviews, looking at competitor products, and talking with our sales teams. Finance has helped us dimension the opportunity. Based on this information we have created a backlog of features our solution needs to provide. These features are expressed as user stories in our agile management tool. You know what a user story is?"

A bit of a supercilious sneer towards Mary was not unnoticed.

Sai intervened, "Mary may know but I don't think I do. Could you give me the elevator speech?"

"A user story is how agile expresses requirements. It's a complete action that provides business value. It is always in the form of a role, an action, and an explanation of value. For example, as Sanjeev the Indian importer, I want to import purchases and rupee sales forecasts from my accounting system, so that I can effectively hedge currency risks. We now have about five hundred user stories, organized into categories we call epics."

"Makes sense," said Sai. "Now what?"

"The next step is to do release planning. We take these stories, estimate them in what we call story points to get their relative size and complexity, then we allocate them to two-week periods we call sprints. Our last sprint zero task is to take the stories for the first sprint and groom them – do more detailing of the requirements so we are ready to enter sprint one."

Mary knew that this is where the team was stuck – in planning the release. Jackson had a method that he was teaching the team, but he wasn't listening to the team. He knew how this should work, it was standard scrum methodology. The team wasn't so sure.

Mary had worked with Jackson to prepare a visual exercise for this step.

"Jackson, lets demonstrate the release planning step. Can we all gather around this side wall?"

Mary had everyone either stand up or take their chairs closer to the side wall where Jackson had prepared the visual display.

Jackson waited for the audience to settle in, and then turned to the wall. He and Mary had set up columns on the wall, outlined in masking tape, with planning headings, and put 2 x 4-inch stickies of user stories into the Backlog column (see Figure 6.4).

Jackson explained what was on the wall as a standard release planning board, replicating in a simpler form the more detailed board he built in the studio.

MEETING TIP

Get people away from the protection/ separation of a big table. Semi-circles of chairs without tables, close to a wall exercise, encourages people to pay attention and interact.

LEADERSHIP TIP

Get others to do the work, especially if you have a "doubter." Leadership means enabling the team. If you do too much yourself as a leader you undermine team effectiveness.

"The Backlog column contains the user stories, which are our requirements and our units of development. Mary asked me to select a few of the user stories we haven't yet slotted to a sprint. I've also handed out the high-level sprint plan we've put together so far."

Mary wanted to highlight an important assumption in the scrum approach, sort of a leading tip.

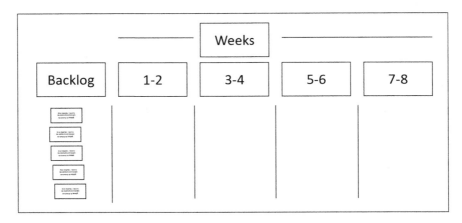

FIGURE 6.4
Scrum release planning board.

"I'd point out that this method assumes that the unit of development is the user story – the backlog is not just a list of requirements, it's the basic pieces of development and testing. That is one of the beauties of this approach, that we develop a full piece of user functionality from beginning to end, piece by piece. Right, Jackson?"

"Exactly, Mary. Agile believes in delivering complete user-valuable pieces frequently. You can see the principles over on the wall," Jackson pointed to the Agile Manifesto excerpts, "including that working software is the primary measure of progress."

"Laudable principles. We need to figure out how they apply to our project. Let's start with the first story, shall we Jackson?" Mary prompted.

Jackson grabbed the first sticky in the Backlog column and read it. "As proprietor of a non-US importer, I want to see on demand a weekly forecast of unhedged payments and receipts so I can make decisions on currency hedging.

"We have talked through the estimate for this and given it five points. That means the team thinks it to be of moderate complexity. We use points to give us a rough idea of relative size – some people think of it in terms or hours or days of work, and some have reference stories for comparison. I don't like the translation to hours, but because we need a reference, I've been coaching our team to think of a point as an ideal day of work.

"The next step in our release planning is to allocate this story to a specific two-week period, based on when it makes logical sense to do given the other stories we are developing."

"This is where I get lost, Jackson," Gary Sunderland (Development Manager) spoke up. "I'm having a hard time knowing when we should do this work without knowing more about the preceding steps. In fact, I have a hard time coming up with points estimates as well. We have lists of user stories, sure, but we don't know which system components are needed for this story, who will develop this, and what they will be able to deliver when. On this one, the effort will depend a lot on what's been built before we begin to work on it, and I don't know that without a more fully elaborated architecture and a more fully laid out plan."

"This was my estimate of five points," said Vivian. "I had to come up with something, so I assumed that the work was just taking data that was already easily available and showing it on a report or a screen. I tried to be sure that there are other user stories that if developed prior to this one would give me that. It's an assumption, and you all know what we say about assumptions."

Gary supported Vivian's concerns by adding, "If I had to guess, the data we need will be in our cloud data store, but to provide just the unhedged transactions we would need to run the cloud data against the CurrX data to see hedge status. Or we would need to build and maintain a combined data store that is accessible to high-performance reporting. There is just no way to know if this is a five or a 50 until we do a much more detailed technical design. And then I see no practical way to finish this across several technical components managed by different teams in two weeks."

Damien Lopez (Test Manager) piled on.

"I agree with Gary on the timing. I know it's not a lot of work to build a report or a screen, whichever this is going to be, but our definition of finish is to complete requirements, design, build, and test in a two-week sprint. If it takes most of the sprint to build this how can I test and debug it in a few days?"

Sai knew that CurrX, which Gary had mentioned, was the foreign exchange management software upon which Pacifica depended. But he didn't know its role in this agile project, so he inquired.

"We don't exactly know yet, Sai," explained Gary. "That is one of my big concerns here. In a usual project we would have got more detail on requirements upfront, done a detailed architecture and design document, run that through the review committee, done statements of work with the vendors involved, had reviews of those with Procurement and Risk Management. In agile we don't want to take so much time and process, so many of these steps are either not being done or will be done in the sprints to come, organized by the user stories."

Jackson leaped in to defend the approach he was teaching.

"Exactly, Gary. We want to avoid the waterfall method, which moves too slowly and has too many handoffs. Instead we take each user story or group of stories, finish requirements and design, build and test, then go on to the next set. Much faster, closer to customers, and a lower risk of building on untested decisions and software."

Mary was sitting back quietly enjoying the dialogue and happy that they had reached the crux of the issue at hand so quickly. She wanted to get to this basic conflict between big upfront requirements/design/planning with which Gary and the existing Pacifica team was comfortable and the agile approach she thought Jackson was naively teaching. She hoped the team could have this debate without engendering too much conflict between the team members.

"Kayla," Mary interjected, "you've done several agile projects, perhaps you have some perspective to add to this debate?"

Kayla knew she had been asked to attend this meeting for this very purpose – to provide some much-needed practical expertise in agile program management. She smiled at the team around her and began.

"Well, my sense is that both Jackson and Gary are right in their comments. We don't want to keep doing the waterfall approach that Gary described, as Jackson argues, but in this case, we can't just do simple scrum without more waterfall-ey groundwork as Gary argues. We need to find the right mixture of planning and methods that works for this particular case.

"My first agile project was a pretty simple one," Kayla continued. "We were building a new website for internal use, mostly content, some connection to existing back-end systems through mostly existing services. It was a small team, developers and testers and user representatives in the room. We used a scrum method much as Jackson is teaching you, and it worked astonishingly well. We moved quickly, suffered few defects, had a close connection to users, and it was a big success.

"My next project was very different. It was about improving battlefield management systems. It involved several different hardware and software vendors and multiple departments in our company and the Air

Force. The simple scrum approach couldn't work – we needed to do a lot more upfront work to be sure everyone knew their roles, had more detailed schedules, more complicated team structures, more definitive architecture documentation and review. We still used a lot of agile concepts but had to build a program structure that worked for all the participants and the complexity and risk we faced.

"Looks to me that this program is somewhere in the middle of these two examples, and that we just need to do some customization of the approach we have begun to make it fit a little better."

At this, Vivian Lockerbee, the developer on Gary's team who had been itching to get to coding but had been frustrated by Jackson's user story elaborations, spoke up. Vivian was smart and cynical, confident and prided herself on rational intellectualism and questioning of authority where it didn't make sense. She wasn't a Jackson fan.

"That makes a lot of sense, Kayla. Hey Jackson, have you ever actually done a project like the one we are doing? Why do you know this is going to work?"

This painted Jackson into a corner. Mary contemplated softening the query but decided to let it go.

Jackson had listened to Kayla intently. He recognized the challenge from Vivian but genuinely wanted the team to succeed, more than he wanted to be recognized as the expert in charge. Mary knew that he valued his connection to the team and had become stressed by the bubbling discontent. Mary thought Jackson could handle this.

"Well, Vivian, I think Gary has a good point, and Kayla's examples are illuminating. I have done a lot of study and observation of scrum projects and have the D&G playbook to follow but have only been on one agile development project personally. It was a website for an auto manufacturer, more like the simpler project Kayla described than the complex one. I'm certainly open to modifying our approach to better fit what we need to do."

Mary recognized the breakthrough that had just happened and took back control of the room. Jackson's Current State exercise was near complete. Mary stood up and walked up to the agenda on the wall. She wanted to summarize this step and lead into the next agenda topic. She didn't trust that others would get it quite right and wanted the efficiency of moving on.

"Let's see if I can summarize Current State for us," pointing to Current State on the wall agenda. "We all understand the good work we have done in understanding customer needs, the competition, and the financial opportunity. Right?" Mary held up her fist. No one responded, and Mary realized this group didn't know fist to five (description in Appendix 2 page 261). She explained and saw Pam Eldridge (Scrum Master) catch the drift and encourage everyone else to do fist to five. Fives all around.

"We understand that before we get a solid release plan – i.e. allocate the user stories to sprint periods, as Jackson laid out – we need some additional upfront planning work, as Gary has advocated." Fist up, fives all around.

"Let's stop here for Current State. We are agreed that we have a solid base of understanding of need and what the product needs to do but need more work to lay out how we execute." Mary summarized this on a flip chart on the wall and put a check mark by Current State on the agenda to indicate completion.

"I'd planned to do an exercise on how we saw the development process proceeding but I don't think that fits given this conclusion on Current State. Let's take a break and come back and talk about what kind of planning we need to do and how we do it. Make sense?"

Fives all around were spontaneous, except for Sai and Melanie Strom (D&G Partner), who were sitting next to each other and had been conferring during the preceding discussion. Mary nodded to Sai who seemed to speak for both of them.

> **LEADERSHIP TIP**
>
> Even the best planned meetings can go off course. Experienced leadership can recognize the flow and adjust with it.

"Great discussion, and I can't disagree with your conclusions," Sai said.

"My concern is that we don't seem to be quite as far along as we've been assuming. What are the cost and time implications of taking a time out now to do more planning? We have some commitments for delivery that we can't afford to miss!"

The room quieted, until Sara Okada (Product Owner) spoke up. Sara had been active on the team in the work up to now laying the customer and market groundwork and converting those to user stories to be developed. She was soft spoken but decisive, passionate about her customers but practical to a fault. She was the business leader of the team to date, so this was in her court.

"I don't know much about technology, as you all know," she said.

"I do know that this team has been working very hard together to get to this point, and I trust that each of you want to do exactly the right thing for our customers and the bank. Obviously, we don't know what doing more planning will do to our schedule, but equally obviously it seems to me that if we don't do as we have agreed the risks to success are much higher. How about we talk through the steps we need to take, and then evaluate impact and figure out how we communicate this to our leadership? Let's not awfulize quite yet."

Melanie nodded her agreement and endorsed the view.

"I'd agree with Sara. We've seen a lot of different implementations of agility in our consulting and coaching. It's not unusual that after a few months of a new project, as the team learns and the scope and environment emerge, we need to take a step back and make some modifications. It doesn't necessarily mean that our initial time and cost estimates are already lost. Let's figure out what we need to, and then let's get it done quickly. We will do whatever we need to do to help."

Sai endorsed Melanie and Sara's statements and the Current State segment came to an end. As the group moved out of their chairs for a recess there was a buzz of conversation and engagement, as they thought about the next segment of the meeting to come about the next steps. Mary in the meantime was already planning how she would help the team structure the upcoming dialogue.

Signposts	The Pacifica team realizes that it needs to take a step back and do some more planning before plowing ahead with the over-simplistic scrum method that Jackson had been teaching. It has a solid base understanding for what is needed but the execution path is unclear.
Leadership Guides	• Extraordinarily well-prepared and executed meetings can help a team understand its position and make good decisions on the path ahead without devolving into personal conflicts. • Focus on facts and fact-based opinions through visual exercises. • Figure out how to get to win–win whenever you can. Mary knew she had to get the team to express their reservations about Jackson and the approach he was teaching, but she wanted to avoid him feeling attacked. A fine balance to walk.

(Continued)

(Cont.)

Coming Up Next	• The first Mary-conducted team meeting ends with an under-standing that the project needs more focus on architecture, stronger involvement of its vendors and internal development partners, a more specific project plan, and a broader team structure able to better include participants who cannot/should not be full time in the studio. • Mary counsels the team to first focus on the "what" they need to build out (i.e. the architecture of the system). • Next, we do some background on architecture for agility, and then we'll see the Pacifica team conduct an architecture simulation meeting.

Section 2

Three Major Frameworks (Architecture, Plan, Team Structure)

- In this section background is provided on three major leadership challenges for large-scale agile projects.
- Establishing superior frameworks for the architecture (including technology and business process together), project plans, and team structures is a core responsibility of facilitative leaders for agility.
- After we learn about each framework in turn, we will see the Pacifica team, increasingly under its own steam, conduct extra-ordinarily well-prepared meetings to efficiently establish alignment around the rigorous frameworks.
- By the end of this section the Pacifica team has the foundation it needs to proceed smartly into full-scale development.

7

Background

Architecture for Agility

WHAT IS THE LEADERSHIP DEMAND FROM ARCHITECTURE?

It's not news that strong architectural work is foundational to enable lean/agile software development to succeed. I won't go deeply into what architecture is, specific patterns, what are good ones, etc. Instead, my focus is on defining the demands on leadership arising in the realm of architecture and on providing guidance and an example for how success can be attained. Throughout we reinforce RAE and use extraordinarily well-planned meetings as critical events and milestones.

In the case of agile delivery projects, the critical elements of architecture effort are:

1. Identification of each piece of hardware and software that must be created or modified in order to deliver, and how they fit together (typically this is called the solution architecture, context diagram, data flow diagram, or similar).
2. A clear map of the business process, from the customers'/users' perspective, to each piece of software or interface.
3. Widespread understanding of the architecture and the interwoven process map by the entire team, so that they are aligned as they make many decisions to come.

I've seen this attempted in very thorough and boring ways that deliver the first element – formal documentation of each systems element and interface – without thoroughly understanding the second – how the technology is to be used. This results in a flawed architecture and incompletely thought-out business processes.

I've seen this attempted in very thorough and boring ways that deliver part of the second element – formal documentation of business process – without thoroughly understanding the first – what are the technology elements and how do they connect. This also results in a flawed architecture and business processes poorly supported by technology.

I've also seen this attempted with both the technical architecture and the business process design, done well albeit separately or silently so not well understood, endorsed, and broadly implemented and evolved by the broader team.

The challenge to leadership is to get ALL of this; 1, 2, and 3!

I'm going to present and show one mechanism that has worked well in variations, and which I haven't yet seen fail. In accordance with our theme of "People over Process," this is not being presented as a cookbook standard – I'm just giving you ideas upon which to base your own planning. Nearly each time we teach another group how to do this kind of work, we see innovation and adaptation to circumstances.

THE ARCHITECTURE SIMULATION

Throughout the last 15 years, the architecture simulation meeting has been a reliably great thing for our projects. Several of these have become legendary (in our own minds, anyhow). I've seen many team members adopt the idea, with audiences of from 10 to 125 people (that was quite a production).

The Simulation delivers and demonstrates a lot of the characteristics of facilitative leadership. It promotes rigorous thinking, gets input and alignment, is efficient … the full RAE. We make the meeting participants do the work, we avoid projecting anything on a screen except for the software itself, we are highly visual, and it's a lot of fun. I'd contrast this with a less RAE architecture walkthrough that comprises some interaction diagrams drawn with an architecture tool within a PowerPoint or word document, projected on a screen, with a boring architect droning on about what the little pictures mean to an audience that doesn't understand and barely cares.

The basic idea is to put on an interactive discovery and learning event. I think of it as a play. There are actors, a director, and an audience, with the roles sometimes fuzzy and changing during the show. The plot is one or more business scenarios, starting with a customer who needs something and ending with the customer needs being met. The actors are representatives of the technical systems and people who execute the scenarios. The director must be a leader with facilitative skills and close to the business process and/or the technologies involved. The audience can be as broad as you like – I typically invite anyone who wants to and is able to come, since one of the primary goals is education and alignment. At a minimum the audience will include as much of the project team as possible, vendor partners, business operations, technology managers, business managers – as much of a cross-section as you can get so input and learning are maximized. You should of course start small to be sure you can do a show before you invite the whole world.

The stage should be a room large enough the audience can gather around one spacious wall. I advise arranging at least the first tier of the audience – the people you expect to participate and learn the most – in a semi-circle of chairs without any tables in front of the target wall. The absence of tables helps the audience pay attention, easily get up and participate, and interact with each other. Remember – this is not a meeting. It's a participative learning event.

I highly recommend against allowing participation if not in the room. With the technologies that I've had access to up to now this just doesn't work – it cripples the room if the focus is to make it work for remote, or remote doesn't get much out of it. Perhaps the latest or next generations of remote meeting technology can overcome this; I hope so! If you do decide to allow remote participation, I'd recommend a wide-angle video view of the room and someone appointed to manage the remote participation by ensuring, as far as possible, they can follow the activities and to take any questions or comments.

Preparing for the Meeting

There are two critical elements of the play we are going to put on: the technical architecture, and the business scenarios we are going to run through it. Figure 7.1 presents a simple example of technical architecture and business scenarios side by side.

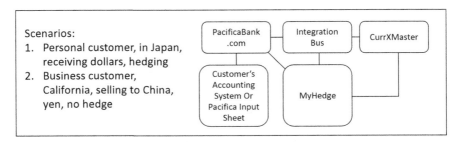

Scenarios:
1. Personal customer, in Japan, receiving dollars, hedging
2. Business customer, California, selling to China, yen, no hedge

PacificaBank.com

Integration Bus

CurrXMaster

Customer's Accounting System Or Pacifica Input Sheet

MyHedge

FIGURE 7.1
Architecture simulation overview.

Figure 7.1 is a highly simplified excerpt from the Pacifica case study. On the right is the technology architecture. Customers access the system through Pacificabank.com in the upper left corner, either directly or through their accounting system or an input sheet upload. Pacificabank .com directly connects to MyHedge, a software application the team is building, and through an integration bus. The integration bus connects dot com and MyHedge to CurrXMaster, a vendor package used by Pacifica Bank. Together these systems and connections make up the business solution.

On the left are two of the scenarios that the business system will support. The first is a personal customer in Japan hedging a future dollar receipt. The second is a business customer in California selling to China and receiving yen, unhedged but monitoring.

The play begins by ensuring that the audience understands the basics of the architecture and the scenarios. Typically, one of the senior technical team members will put the architecture together, a key preparatory task. The architecture will be on the wall in some way – from labelled colored paper with masking tape as interfaces, to writing on a whiteboard wall, to physical stations occupied by the actors arrayed from left to right along the wall. The best specific method for your meeting depends on many factors, such as level of detail, maturity of group in doing this kind of meeting, participant preferences, time available, and more. You can't go too far wrong. Have the architect explain what is up on the wall.

Another team member will be tasked with assembling the business scenarios. These can start with personas, archetypal system users often designed in the early stages of initiative planning. The goal is to assemble a few simple scenarios to get the basic flows understood, and then vary by adding new dimensions and complexity to flesh out the boundaries. The Product Owner, the User Experience Manager, or other sales/marketing/operations staff should be tasked with this.

In this example we might start with the simplest case. In Scenario 2, the Californian exporter might begin by logging into Pacificabank.com, then navigate to MyHedge via marketing pages, and sign up for a free trial. The exporter might then upload a file from an accounting system of accounts receivable, validate that the data has been accepted accurately, and select a future interest rate scenario. The output could be a report showing the US dollar value of the yen sales under various future interest rate paths. As the team walks through this scenario, it would follow the data from component to component in the architecture, describing what happens at each step. The result is a lot of inquiry and understanding.

As in any RAE event you need to be clear on what will come out of the simulation. You should appoint someone to update the architecture documentation and someone to fine tune the scenarios for next use. Another team member should keep notes on issues or parking lot items, both on a visible wall chart and documented for the team to address later.

The last item is to make the invitee list and prepare for it to grow.

Variations

Architecture simulations can be adapted and scaled to meet a variety of needs and can succeed even if many of the tenants elucidated here are not well done. Here are two examples, starting with a quickly prepared two-system exploratory event, to a highly prepared, dozens-of-systems confirmation event. Following which we will see Pacifica do their own simulation.

I recently participated in a simulation designed to tease out the relationship between a mobile app being developed by one company and a responsive web application built by another. The traditional agile method might have been to establish personas and journey maps and user stories, allocate them to sprints, and then have the sprint teams figure it out. This seems like just another way to do waterfall or big requirements for upfront development. Instead, some of the team leaders chose to give an architecture simulation a shot.

The meeting plan was quite simple. Get representatives of the two systems together, have a few scenarios, have the mobile app up on one screen and the responsive web up next to it on another. Then run the scenarios and figure out integration points and methods. Prepare well – have the architects and engineers and product managers on both sides bring along diagrams and roadmaps. Wrap up the meeting with next steps.

However, much went wrong in the execution.

- The room was appallingly ill-suited. We wound up in the board-room, which had big luxurious chairs surrounding a large fixed table taking up most of the room.
- There was only one screen, so it was impossible to see the systems side by side.
- The seeding architecture diagram was only available to partici-pants on paper and was overly broad and technical in nature. It included too many superfluous systems and was quite inaccessible to most of the participants.
- The user experience and product management team members from the mobile app couldn't or wouldn't travel to the meeting and participated poorly from afar.
- We couldn't figure out how to project the mobile app at all, so we had to use PowerPoint screenshots and hand around the iPhones of participants in the room who had installed it.
- The demo environment for the responsive web system was barely functioning.
- The agenda was overloaded with background presentations prior to getting into the heart of the simulation.

Nevertheless, we wound up with a good outcome: a rigorous and efficient alignment of what had been quite a fractious assemblage of goals and proposals. Some of the participants had enough background in architecture simulations to adjust the agenda in real time and to keep bringing the group back to looking at the actual systems in the context of the scenarios. That simple framework enabled input from many of the participants and the team made dramatic progress that got the project off to an excellent start.

At the other end of the architecture simulation spectrum is one done part-way through a two-year project aimed at finding any remaining gaps to reduce risk of failure. This project touched more than a dozen systems and business units, and when live would present modest business process changes for thousands of users and dramatic changes for a hundred or more. One of the leaders of this project had recently joined the team and had seen large-scale architecture simulations succeed in ensuring rigorous efficient alignment in an earlier program. She reached out for coaching from prior colleagues, convinced her leadership to give it a shot, and devoted weeks of preparation and practice to the event. The elements she focused on included:

- The right room and room arrangement.
- The systems architecture on the wall.
- The scenarios.
- Representatives of each major system recruited well in advance, joined together along with a few others on the simulation sponsoring team. This group of about 10 took ownership together for the success of the event. They practiced!
- The audience. Because there were dozens of systems involved (about 15 specifically on the architecture diagram on the wall, with some represented in groups), multiple business units, multiple staff groups (legal, compliance, risk, reporting, training and communication, etc.), and multiple levels of management in each unit, the invitee list quickly grew from 25 to 50 to 100. As the word spread the audience on day one wound up at 125 people.
- Senior management support. This kind of event had never been held in this organization and getting over 100 people together for this detailed a run-through of technology and business process together was not part of the culture. In this case, senior leaders were worried about a potential failure they could not afford and saw what they didn't know as the biggest risk. Further, they realized that the good work done to date meant that the remaining failure points were most likely "in-between" – in the system handoffs, the business process integration with system behavior, the accuracy and completeness of the training materials, and the quality of user and production support.
- Consideration for remote attendees. The team set up a webcam and a chat room with an attendant to help remote attendees participate. The rules were very clear: remote folks can view and ask questions but can't have a lead role. Lead players had to be on stage. Senior leaders provided funds for travel so that there would

be no financial excuse not to be there. As an aside, even with such attention given to the remote audience, feedback was clear that remote participation did not work well at all.
- Dedicated time. The meeting was planned for two full days and regular meetings that could conflict were cancelled or rescheduled.

There was a lot of skepticism going into this grand simulation, but also a lot of enthusiasm. Coming out of it was a consensus that it was one of the best meetings people had ever attended, and two compelling conclusions: there were a relatively small number of gaps identified that could otherwise have de-railed the implementation, and that by and large the effort was on a good track to succeed. In talking with participants, it was clear that while the formal documented issues raised were modest, the number of improvements and action items taken away by the audience was quite large. Perhaps more compelling but less measurable was the greater broad understanding of the ecosystem and improved decisions all around for the remainder of the project. After the project went live and the core team did a formal retrospective, this simulation was highlighted as a major contributor to its success.

Enough of examples. Let's see a simulation in (fictional) real life, at Pacifica Bank.

Signposts	Solid alignment around the system architecture, the business process, and their interaction is foundational to enable agility.
Leadership Guides	The architecture simulation event is a proven mechanism to discover and build alignment around architecture. It can be used in many situations and at various stages of a project. It puts the focus on the software and the related business processes in a powerful way.
Coming Up Next	Pacifica Bank does its architecture simulation.

8

Pacifica

The Architecture Simulation Meeting

SETUP

Mary had the nanny come over early the morning of the simulation so she could arrive in time to make sure the setup was right. She had worked intensively with Pam Eldrige (Scrum Master), a natural hostess/organizer; David Phillips (Architect) who was responsible for arranging the system flows; and Judy Grossman (UX Manager) who would be representing the customers and bringing the scenarios to run through David's systems. As she walked into the studio at 8:15 AM and surveyed the expanse she was pleased with what she saw, as the three developing leaders buzzed around the room preparing for the meeting to start at 9:30 AM.

The back wall of the studio, under the deck containing the huddle rooms, was set up with the systems flow (see Figure 8.1). It was surrounded by a semi-circle of what looked to be about 50 chairs, with 20 feet or so between the first row and the wall, leaving room for people to easily get up, talk, and draw. On either side of the systems flow on the wall were large portable whiteboards and flip charts. Top level, looking good!

Mary said good morning to David, at the wall arranging and taping. David turned around and smiled at Mary and asked if she would take a look.

David and Mary had talked over how they wanted to do this first simulation and had agreed to a hybrid of a wall exercise and a full-blown role-playing exercise. Pam had thrown in her advice as well, thinking that since the team was just starting into this more intense method of working together, they should begin modestly. Mary asked

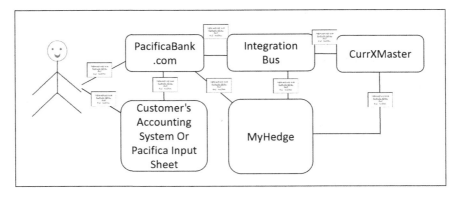

FIGURE 8.1
Pacifica architecture simulation wall.

David to give her a tour and explain how the exercise was coming together.

"Pam and I thought that we'd put the architecture up on this wall, with the customer facing systems on the far left, flowing through the back-end systems ending on the right. Each system has its own sheet of colored paper – the colors really don't mean anything, but Pam likes the color. Interaction between systems is represented by masking tape, and I put these stickies up on the tape with a little information on the interaction such as real time or batch, what mechanism, what transformations or logic we put in the middle. We have a basic idea of what the elements involved are but have some rather significant unknowns on how it all fits together."

"I love the stickies on the interfaces, David. Often folks just assume away the connections among the systems, so that's a helpful tool at this stage." Mary was tickled with David's progress – he was already innovating and making the exercise his. "How are we going to involve the team in the transaction simulations?"

"Pam!" shouted David. "Could you come over here and show Mary the name tags?"

Pam was just finishing up helping the caterers set up the coffee table. She stopped by a table against the side wall and brought over a colorful armful.

"You said to get people up and involved and to make things visible," Pam started. "So, David and I thought we would have a person represent each system on the diagram and have them come up and explain system behavior when it's their turn. I wrote the name of each system on a piece of colored paper and stapled the paper to these plastic lei I had at home."

"Perfect!" Mary excitedly exclaimed. "Who is going to be taking notes and updating the architecture and issues list?"

"Vivian," responded David. "She is going to put up a chart over there," pointing to the side wall, "and then update the draft architecture document we've started. Gary and I also asked Kayla, since she is new to this realm, to assume the role of Grand Inquisitor – wherever something isn't clear and no one else is digging in, she is going to ask the tough questions."

"Sounds good. Nice to get developers deeply engaged in the architecture development, and I love the role for Kayla. Rigor! Carry on … I need to check in with Judy."

Judy Grossman was sitting quietly arranging some materials at a small table on the left side of the simulation wall. Mary dragged a chair over from the semi-circle fronting the wall, sat down next to Judy at the table, and asked the User Experience Manager about her plans for the day.

"I'm going to be representing our users as they try to get their work done on our new solution. I thought I'd just use the personas we already have. I'll start each scenario out by explaining who I am and what I want and why. I've got a different hat for each persona to make things visible and avoid confusion – I hope no one gets offended!"

Mary had noticed a small box of what looked like clothing, but she now saw was colorful hats. "If you explain the purpose and mention your concern with stereotyping it'll probably be OK. But be careful.

"How do you plan to walk through the scenarios?" Mary further inquired. "I brought this old PC, which I'll use when I pretend to work with my accounting system and access PacificaBank.com. To make things more visual, I have folders with big letters on them representing the information I'll put into the systems and what I will get out. Here on the wall," pointing just left of the stick figure of the customer, "I'll keep track of what I'm doing so everyone can see."

"I'm a little nervous actually, Mary," Judy confessed. "Sara helped me think through the scenarios and promised to come up and help if I need it, so that is a bit of a comfort." Sara Okada (Product Owner) was more deeply knowledgeable of the details of the product, but Mary had advised the team to have Judy be the customer. This would give Sara more freedom during the meeting to be involved in each step as it unfolded.

"Looks to me like you are all ready to go," Mary said, as she took her leave of Judy and headed over to the coffee table as the players and audience began to filter in.

KICK OFF

Pam had been excited to try to facilitate the simulation and Mary had been pleased that the D&G hadn't tried to take it over. The agile playbook aka scrum that D&G was teaching didn't have a page for architecture/process simulation. Jackson Maxim (Agile Coach) had by now welcomed Mary's help and even Melanie Strom (D&G Partner) had turned positive on Mary's contributions. Melanie had even felt Mary out obliquely on her potential interest to join D&G permanently. In all, the group gathered together to simulate seemed unified and anxious to proceed.

Mary thought Pam a real gift to the team. Somewhere between motherly and grandmotherly, clever, and a super team builder. Pam was the cheerleader, the arranger of birthday parties, the loyal concerned friend. Not highly technical but certainly experienced enough, especially able to think about users based on years in support call centers. She'd learned scrum and got herself certified, not a deep thinker or innovator but a steady harmonizing get-things-done-for-the-team practical leader. Pam had been through workshop facilitation training a few times in her career and had done quite a lot of this kind of work. Mary only needed to add some focus and sharpening, and Pam was ready to rock.

With Mary's prodding and guidance, Pam had put together good written and visual agendas and was ready to go as 9:30 came. She gave the group a 5 minute warning that the meeting was about to start, while David and Judy helped her herd the attendees into the chairs at the front of the room.

Pam kicked the meeting off by welcoming everyone and asking Sai to say a few words.

Sai looked around the room and echoed Pam's greeting. Then he began setting the stage.

"We are a few months into this new agile world," he began.

"As you all know, Pacifica is a leader in trade finance and capital markets throughout the Pacific basin, and we have been for more than 100 years. There has been an amazing amount of change and progress in that time ... just think, forty years ago China and the western countries didn't even have diplomatic relations. The internet was just a dream at a few universities, and the idea of walking around with a powerful combination computer/communicator in our pockets had just been introduced by Captain Kirk."

Sai reached into his pocket and pulled out his iPhone.

"Can you imagine if someone told our parents in 1970 that a poor kid born in Bombay – that would be me – could be leading a business based in San Diego trying to help Japanese and Vietnamese entrepreneurs by building software for this communicator that was made in China? No one would believe it.

MEETING TIP

Having a very senior leader kick off an important event is often appreciated by both the team members and the executive. Team members tend to crave connection and recognition, while executives often have more interest in the work of teams like this than you might expect. Oftentimes executives invited to kick things off for a few minutes wind up staying!

"But here we are, and the pace of change isn't going to slow down. With all your help, Pacifica aims to continue to prosper and serve our customers for another hundred years. To do that, we need to change – get closer to our customers, be more digital, move more quickly.

"One way we are trying to get better is this Agile Studio, and this project. It's our first major project to try this new agile approach. We've chosen a critically important market for us where we believe we have a great opportunity, and brought on D&G," nodding towards Melanie Strom and Jackson Maxim, "to help us learn how to do this. Most of you have also met Mary O'Connell," nodding towards Mary, "who has a lot of experience with agility. And we've assembled some of our

LEADERSHIP TIP

Bring vendor partners into your agile projects as soon as you know they will be an important part of the solution. Creating win–win partnerships can entice partners to assign their best people, modify investment plans, and accelerate development of features you may need.

best people at Pacifica, from every major function we believe needs to be engaged and put you all into this beautiful new collaboration space."

"Today I want to welcome some new participants as well. I've learned that one of the agile values is Partnerships over Contracts, and that we need to learn to deal with our vendors in a much more collaborative win–win way.

"We believe that TradeX, and the system we already use to manage our foreign exchange CurrXMaster, will be a central part of our solution. Rather than doing an RFP or something formal through procurement, we asked some of their leadership to join us here, early in the solution design. Ivan, can you stand up? Welcome to the team.

"Enough from me. Let's get going! Pam?"

Sai sat down, and Pam took the room back.

"As long as we are talking about participants let's do introductions. We'll go around the room. Each of you please stand, say your name, organization, and role, and any specific concerns or goals you have for today. Ivan, since Sai just pointed you out, would you mind starting?"

"I'm Ivan Ponomarenko, product engineer at TradeX. My goal is to understand what you want to do and figure out how we can help. I'll be able to tell you what CurrXMaster can do now, and if we find things we need to build I'll bring them back and guide consideration of product enhancements. I work regularly with Dexter McDonald here at Pacifica."

"That's me," said Dexter as he stood up. "I am a solution engineer on the financial transactions team. My job is to lead the design and planning for CurrXMaster and any other financial systems of ours you may need. Thanks for including us in this event."

Over the next 15 minutes the remaining participants made similarly brief introductions. Mary noted that two new key players who had not been in the studio until now had joined: Lilian Kim, a senior systems analyst from the PacificaBank.com team, and Yong Zhou, an architect for the Pacifica shared services division that managed the enterprise service bus through which most system-to-system integrations flowed. They were sitting with Gary Sunderland (Developer), who had recruited these peers to attend.

As introductions were completed, Pam checked off Introductions on the wall agenda and flipped her chart to Objectives. The Objectives chart had a diagram Pam had worked hard to draw up – she tried several formats before settling on the one shown in Figure 8.2.

"We have three inter-related goals for today. The first is to examine and refine our systems architecture. This is a diagram of each technology element that will be part of our solution, what each element does as

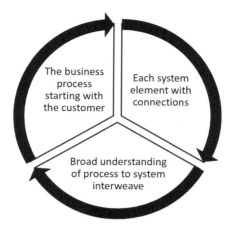

FIGURE 8.2
Objectives of architecture simulation.

part of the solution, what changes we need to make in those elements to get to our solution, and how the elements are connected. David has a starting point up on the wall," Pam said as she pointed to the architecture diagram. "The second goal is to understand the business processes from the point of view of our customers. Sara and Judy have put together a number of scenarios, ranging from very simple to modestly complex, that they will share with us.

"The final goal is that everyone in this room walks out with good understanding of how the business processes Sara and Judy will share interweave with the technology system that David has articulated, as we modify it.

"This broad understanding will be documented and will serve as the foundation for our next steps in building out the project plan and schedule. Once we know what we are building and how it will serve our customers, we will have much more of the information we need to lay out the tasks to build and deploy the solution and get a good schedule to completion."

A brief conversation ensued, with no significant change or challenge to the objectives. Mary pitched in a brief story of how this meeting format had evolved and what she expected out of it, giving an added dose of confidence to the attendees. While the meeting plan was novel to everyone in the room with the sole exception of Mary, Pam's explanation made sense, especially in view of David's diagram on the wall. Pam sensed the room was ready to get into the meat of the day so she returned to the agenda on the wall, checked off Confirm Objectives, and pointed to the remaining agenda items.

"Our primary exercise is the simulation you see up on the wall. First David will explain the technical architecture as we currently understand it. We plan on spending about a half hour on that but let's see how it goes. We want everyone to understand the various pieces and connections at least at a high level. Then Judy will describe the business scenarios we are going to simulate. Next we'll take a break.

"When we come back from break, we will start on scenario one. We'll start from the customer perspective, see how the customer engages with our systems, and follow the systems' flow from beginning to end. As we do that each of you should test the boundaries, be sure you understand how things fit together, and ask plenty of questions. Once we finish the first scenario, we'll do variations of increasing complexity to help us flesh out unknowns and complexities.

"At the end we will leave some time to wrap up what we've learned and be sure we have a good list of issues we need to follow up and to set next steps – which we are currently thinking will be release planning.

"Any final questions before we start? If you can hold questions, please do – I know it's hard to imagine how this is going to work but let's start and see how it goes." The audience was willing, so Pam had David come up and start to explain the architecture.

REVIEWING THE GOING-IN ARCHITECTURE

As David walked up to the front of the room, Mary reflected on how well David fit the bill for this team. Around 40, Mary guessed, an engineer out of Berkeley followed by a decade at Microsoft in Seattle, ending up Microsoft Money as it was discontinued and absorbed by Quicken. At a crossroad, David had moved back south with his young family and settled at Pacifica to what he hoped would be a stable and not overly demanding work situation. Strong technical skills, a good teacher and mentor, there was something of a dreamy professor in him. He seemed quite at home at the board in front of this group.

David began at the far left of the wall with the customer (see Figure 8.1). "Our customer is a small business or wealthy individual customer who has revenue and expense in multiple currencies. The cash flows are big enough that potential fluctuations in exchange rate might matter, but not so large that it's worth spending a lot of money to manage them. These customers can't afford staff devoted to managing currency risk, and today have no cost-effective means to understand or limit risk besides seat-of-the-pants calculations or guessing." Pointing to the box labelled "Customer's Accounting System Or Pacifica Input Sheet," David continued. "Most of these customers have a multi-currency accounting system. There are about a dozen of these systems that comprise 90% of the market. We do not plan on competing with these – instead we plan on integrating with them. They can all export data – we haven't finished our analysis of how that integration will work, but we know we can make it work for two or three so far and are optimistic we could do for most or all. We do know that some potential customers don't have one of these tools but could still benefit from our hedge support, so we are thinking we'll do spreadsheet input and web page mechanisms as well.

"So, the source of the data for the hedge request is the customer's data file. The customer will provide that data to us through Pacifica Bank.com. We will need to create some marketing/introduction pages on the public site, and then use our existing security mechanisms to sign the customer in. The customer then uploads the data and gets back some options to reduce currency fluctuation risk. The customer needs to be able to update his or her projected cash flows as things change and see options to adjust the hedge. The options could be very simple, such as 'reduce risk by half' or more complicated, whatever we all decide in the studio. Finally, our customer must be able to view the status of hedges, fees, and other status information on demand and get the kind of notices he or she wants. It has to be very simple for the customer, but to make it simple we have a lot of complexity to deal with."

Lilian Kim (PacificaBank.com Analyst) had a question. "David, how much of this do you expect dot com to build? Is it just the marketing pages and the authentication, or do I need to provide the views into the hedging information or more?"

"Right now, we are thinking that you do the minimum – just get the customer and data securely to us and hand off control. You see the MyHedge box in the middle? We plan on this being a new system that Gary's team will be building. We tentatively think MyHedge will have a user interface layer that we'll make fit into PacificaBank.com, a logic layer for rules and calculations, and a data layer to persist information about the customer and positions that don't fit well into CurrXMaster."

"Where is MyHedge going to be?" followed up Lilian. "Any unusual single sign on or data security issues to figure out?"

"We are planning on doing MyHedge in the public cloud," Gary Sunderland, the development manager, explained from the back row. "We need it to be accessible throughout the world, always up, and able to scale if this thing takes off. Vivian and a few of our developers have been playing around with tools and environments and working with Information Security to start the certification process."

"That's going to add some complexity," Lilian said. "We haven't done any tight integrations with cloud-hosted systems."

"It shouldn't be too difficult," David said reassuringly. "Uses pretty standard methods. But it's a good point, Lilian, we'll need to plan some time to do security reviews and take some risk out by connecting up soon." While they were on the topic, David decided to dig deeper into MyHedge. "Gary," he said, "can you give a few more details on your thinking about MyHedge?"

When Gary started to explain from his seat in the second row of chairs, Pam interrupted, got him up at the front of room and put his lei name tag around his neck. Gary sheepishly accepted it and stood in before the architecture diagram on the wall and began his exposition.

"Our challenge is to fill the gaps between our current online and back-end systems and our currency hedge system, CurrXMaster. CurrX can create and manage groups of transactions, model hedge strategies, take trade orders, execute and settle trades. However, it doesn't provide a user and data interface our target market can easily use, manage notifications adequately, or productize this kind of low-volume hedge management. Think of MyHedge as the system that our customers will access and that will then use the capabilities of CurrX to provide a portfolio of transaction hedging services."

Kayla Robinson, who had been given the designated role of inquisitor and asked to be on the lookout for clarifications to make, jumped in here for the first time. "Gary, I see on the diagram an integration bus in between you and Lilian. What do you think that is going to do versus what you will do?"

"I'm not entirely sure," responded Gary. "The integration environment has some capabilities we will probably want to use, such as data format translations and simultaneously updating me and Ivan," looking over at Ivan representing CurrX. "We've had some initial discussion with Yong on this and we want to dig deeper today and tomorrow. Yong can do quite a bit, but the cloud environment has its own overlapping capability. It's not obvious which system should do which functions."

"Exactly why we are doing this simulation," smiled Pam. "What's next, David?"

David stepped back into the middle of the circle and continued down the path. "So we have the Pacifica.com role, and the basics of MyHedge – let's turn to CurrXMaster. Dexter, could you and Ivan come up?"

Dexter McDonald was what David called an old mainframe dude. He was practical and expert in what he did, on the tail end of a long and productive career in the first generation of business computing. He was evidently enjoying being on this new kind of team.

Dexter and Ivan got their name tags on, at Pam's prodding, and walked into the spotlight at the center of the semi-circle in front of the architecture diagram.

Dexter began the exposition.

"We run CurrX in-house on the mainframe. Our version is based on the CurrX system Ivan's team builds and supports but has quite a bit of customization that we've done over many years. We try to stay current with Ivan but in practice diverge a lot. For this project, we want to understand what we need and see if the needs are applicable to the broader CurrX customer base, if CurrX might want to build in a way we can use.

"Think of CurrX as our central facility for managing currencies. It can do any kind of trade we need here, from a simple conversion to calculating risks for large sets of future transactions, including modelling based on different forecast scenarios. It has real-time access to exchange rates and option rates and is connected to our customer and risk and transaction accounting systems for trade execution. Input can come from its green screens, which are good for regular users since they are efficient, fast, and reliable if not terribly user-friendly, or from batch or real-time interfaces.

"We just need to understand what this project is trying to do and then plan out how we could support your project given everything else on our plate. Our initial thought is that we are missing a few elements and would either have to build them or see if TradeX is interested. Ivan?"

Dexter gestured to Ivan Ponomarenko to take over the presentation. Ivan was one of the few attendees in a business suit, having flown in from New York and accustomed to the more formal atmospherics in the financial capitals. Ivan still had remnants of his native Russian language but spoke slowly and clearly and was understandable enough.

"This is a market we have been considering for some time," Ivan began. "We have dozens of installations of CurrX around the world. The system is a comprehensive multi-currency manager from trades to hedges. This idea, managing modest value transaction portfolios, is a new one that we find very interesting.

"We think we have most of the foundational elements you will need but will need to add some specific elements to make the business process work effectively for customers with much less financial expertise and less attention span for currency matters. We also may have some adjustments to make to manage a much larger number of small portfolios instead of a small number of large portfolios."

Kayla was by now relishing her role as Inquisitor and the audience had warmed to it as well. The audience seemed to be expecting Kayla to jump in as she did. "Ivan," she asked, "what do you think those elements are? Are these on your roadmap? What is the timing?"

Pam slowed Kayla down, and managed the presentation of Kayla and other's questions one by one. The dialogue around CurrX, and then back downstream to MyHedge, the Integration Bus, Pacifica .com, and the customer's accounting system continued for more than 2 hours, running well past the planned half-hour scheduled slot. As the degree of overrun became evident, Mary, Pam, and David conferred and agreed that this was valuable use of time and they let it run its course until the lunch break.

Over sandwiches and chips, Mary gathered Pam, David, Judy, Sara, Gary, and Kayla to take a measure of the morning and make course adjustments. They all agreed that the decision to let the group go deeper into the going-in architecture had been valuable, getting the assembled horde to a place of solid understanding of the technology base at hand and ready to start the business scenarios. They had less time to do the variations of scenarios Judy had prepared but Mary assured them this was not unusual. Getting even one scenario completed at this level of detail was of great value.

Judy and Sara hurried through their lunch so they could do final scenario preparation. The buzz within the small groups gathered around the patio as they finished their repast was promising.

RUNNING SCENARIOS

Pam had rounded up the troops and was quieting them down and getting them into their chairs in the semi-circle surrounding the architecture wall. The wall was now plastered with scattered sticky notes and masking tape indicating elaborations to the going-in model, and two flip pages of issues and questions were posted to the right.

MEETING TIP

You have no choice but to plan an agenda based on your best guess on timing. Compelling activities in complex areas tend to go long. As meeting facilitator you need to decide on value and adjust time and topics as best fits. Bring others in on the call. Don't be afraid to go deep even if it takes scheduling another session to deal with items that were late on the agenda.

Pam started at the wall agenda, reviewing the completed segments from the morning and confirming that the team had now indeed sufficiently completed the architecture overview. Per the plan at lunch, she asked Mary to give some perspective and kick off the next segment.

"First, I'd like to do a quick check on your response to this morning's activity. We can use a response tool I call fist to five. I'll state something, and if you agree strongly you hold up five fingers. If you disagree strongly, you hold up your fist. One finger is some agreement, and so on. For example, if I were to say 'that was a great lunch,' and you really hated it, hold up your fist. If you loved lunch, hold up five fingers. Got it?"

Nods, but not what Mary wanted.

"So how was lunch? Give me fist to five on it." Up went many hands with five fingers, some enthusiastically, some reluctantly.

"OK. First thing to check: 'I've now got a good understanding of the technologies we are working with.' Fist to five."

Mary looked around the room at dozens of hands, mostly with three to five fingers. She noticed that Sai, who had to leave for the last half of the morning but had rejoined for lunch, had just a single finger up.

"Sai, I see you have a single finger up. Do we need to get you some remedial education or are you OK?" she ribbed him.

"I'm just being honest, Mary. I'm not much use at technology even though our business is so dependent on it, and I'm a little embarrassed actually. I can see how intertwined the product offering and our internal business processes are with these systems and would like to learn more. Could someone maybe give me a tutorial?"

Gary happily volunteered to try to get on Sai's schedule, so Mary continued.

"We spent enough time on the architecture, let's get to the scenarios. Fist to five."

Mary called on some of the outliers, learned that some thought we needed to go deeper, while others thought too much time was spent.

"In any team like this," Mary summed up, "we'll never get it exactly right. My take as an observer is that we are just where we need to be to start the scenario work, even though it took longer than we had slotted. We have several systems involved, with some complexity in versions and deployments and expected roles, and we haven't as a group tried to put these together productively in the past. One other observation I'll make is that this intense collaboration among technical, marketing, operations, and other functions is new to Pacifica, so a lot of you are being exposed to sausage-making in a new way. I can reassure you that the investment of time in learning the mechanics will be valuable. But let's prove it – let's go on to the first scenario."

Mary stopped at the wall agenda and checked off Architecture Overview (Figure 8.3) before returning to her observation post in the back of the semi-circle. While Mary moved away Judy and Sara assumed their spots in the front to the left of the architecture wall. Judy explained her role as User Experience Manager and Sara's as Product Owner and provided some insights to the work that the studio had done in defining personas, journey maps, and user stories. Now, Judy explained, we would run through as many scenarios as we could and watch how the user tasks and the system behavior would interact.

"The first scenario we are doing is our simplest, base case. I'm going to play Sanjeev. I live in Delhi in India, where I import and distribute building hardware like door handles, countertops, hinges, garage door openers, and such. I buy the materials from vendors in China, Japan, Europe, and America, denominated in their currencies. I sell to distributors and chain stores in India in rupees. My company has thirty

✓ Introductions
✓ Confirm Our Objective
o Architecture Overview – David
o Understand Business Scenarios
o Simulate Business Scenarios
o Wrap Up: Review Issues, Parking Lots
o Set Next Steps

FIGURE 8.3
Wall agenda for architecture simulation.

employees and might be considered a speciality importer – we handle certain brands in certain categories, some special orders, some to warehouses here in India to support more rapid delivery and some for direct shipment to our customers.

"I am the financial manager of the company, and report directly to the owner. I have two assistant accountants who keep track of inventory and orders and manage accounts receivable and payables. I manage the banking accounts myself. We have a multi-user multi-currency accounting system, one of the popular global brands. The system is on this PC, which I also use to access our banks online."

Judy pointed to the old PC she had set up on the table just to the left of the architecture wall.

Sara took over from here. "Sanjeev banks with a local Indian bank and with three international banks, one in Europe, one in China, and Pacifica. He is happy with Pacifica services and typically prefers our service and rates for currency conversions and more complex trade finance.

"As the business has grown, Sanjeev's boss has become increasingly concerned with currency fluctuations and mismatches. Sanjeev has been simply taking currency risks. For example, he will order 10,000 euros of countertops from Italy, denominated in Euros and payable net 60 with a letter of credit from Pacifica guaranteeing payment, and eventually sell the material weeks to months later in rupees. If the Euro/rupee exchange rate moves, his profit can be doubled or wiped out.

"Sanjeev has been offered currency options, forward rates, and other tools that larger companies have been using for years but he gets lost in the complexity. Furthermore, the costs to use such tools on his smallish transactions are too high. When the Pacifica trade finance rep mentioned our new product, Sanjeev was excited to give it a try. Sanjeev's boss was recently angry with him for some losses when the Euro appreciated against the rupee, wiping out the profit on a deal he'd worked hard on."

Judy took back the reins. "This scenario begins with Sanjeev uploading his projected cash flows by currency, choosing a hedging strategy and buying it. We are assuming that he has already decided to use the services and is set up on the system, has created and tested his system output, and decided on what cash flows he wants to hedge. All of these are included in other scenarios we will do later if we have time, but we want to start on the core of the solution."

Pam walked to the center of the semi-circle and checked for readiness to proceed. There were a few details inquired after and then Pam gave the nod back to Judy to begin.

"The first thing I do is log into my computer," Judy explained as she sat in front of her prop and pretended to log in, "and create the export file from my accounting system. I place it anywhere accessible to me, on a directory in my PC or my network, it really doesn't matter."

"What's the file format, Sanjeev?" asked Kayla the Inquisitor. "We need to know what formats we will need to deal with."

Sara (Product Owner) fielded this one.

"We are thinking that we need to keep Sanjeev away from anything technical. Several of the popular accounting systems have export capabilities, some proprietary and some in a common format popularized early by one of them. We are thinking we start by supporting four options – three of them from these vendors, and one that we will create and support."

Gary took off from there.

"Kayla, we've looked at these formats and they aren't all that different from each other. They all contain a transaction set of inflows and outflows, each with an amount, a date, and a currency. Some contain information about the company and the system, some have other information such as payee or customer. The formats range from a simple comma delimited file to an Excel file. I don't see anything too complicated about creating a file translation service within MyHedge to bring everything to a common data structure."

"We could also do that in the Integration Bus," interjected Yong.

"Let's wait till we get there," Pam tried to keep order and pointed back at Judy. "Sanjeev hasn't even uploaded his file yet."

Judy continued as Sanjeev, "Anyhow, now I have my file, so I log into my account at PacificaBank.com. I navigate to MyHedge ... "

Here Pam interrupted and asked Lilian Kim (PacificaBank.com Analyst) to don her lei label and come up to the front. As she ambled up, Vivian turned on the projector and brought up the PacificaBank.com landing page so everyone could follow along.

"Lilian," Pam asked, "could you walk us through how this will work?"

"Sure, Pam. Sanjeev gets to our front door and inputs his ID and password here. He goes through authentication and then the next screen shows him his customized menu panel based on our customer information system and authorization database. Vivian, can you go to the next slide please?"

> **MEETING TIP**
>
> Keep the activity on track. Some minor wandering is OK but the participants need to know where they are on the way to the outcomes promised at all times.

Lilian had mocked up Sanjeev's experience as best she could from what little she knew. The mock-up, while preliminary, nevertheless helped everyone visualize the flow and the choices Sanjeev would have.

"Sanjeev, through his ID/password, is linked to his company and the services they have with the bank. Our security system is role-based, so we know for each service what Sanjeev is authorized to do plus some of his preferences. Here you can see that Sanjeev is presented with a screen in English, and he has several options. He can do personal banking, since he is a customer himself and not just on behalf of his company, or he can do one of several things for his company. He can see account balances and drill into transactions, he can trade currency, he can buy investments, he can do trade finance. And now we are adding a new menu item, Manage Currency Exposure. If Sanjeev selects this menu item, we hand off to Gary."

"That looks generally right, Lilian," Sara remarked. "We haven't decided what to call the product yet, or what the menu item should say. We think we would like to show some information on outstanding hedges on this page, kind of like we show deposit account balances and outstanding letters of credit. We'd also like to be in English, Hindi, Mandarin, Japanese, and Korean to start. When we are successful, we will want to add more of the Indian regional languages like Kannada and Tamil.

"No problem on the names or the languages to start, we don't care what you want the menu to say. We'll have to dig deeper into any preview information, we need to have very fast access to that, we might need to get pre-feed into our cache so we can do that. Vivian, can you write that up on the issues list so we work through that with Sara and Gary and Dexter?" Lilian asked.

"One other item to consider is whether we need more than one role for the project. If we do, we would have to set that up in our permissions and roles service and pass that along to MyHedge."

"Let's assume we will have more than one role, Lilian," Sara agreed. "Sanjeev and perhaps his assistants will be able to update hedge information, while probably only Sanjeev can buy a new hedge. The business owner will probably want to be able to see hedge status and maybe hedge purchase options. I'd guess there might be three or four roles a user could have, with maybe a dozen potential permissions."

"OK, I'll plan on that," Lilian stated. She looked over to Vivian's list to be sure she had captured the need accurately and went on to discuss how she would call MyHedge with Gary. Kayla interjected a mix of naïve and insightful questions, joined by audience members periodically, as they worked through the handoff to MyHedge and reviewed a mock-up of what the landing page in MyHedge might look like. As that conversation concluded, Judy took back the focus.

"Back to Sanjeev. Now I'm logged in to MyHedge, and I select the menu item Model a New Hedge. I get to a screen that looks something like – can you put it up, Vivian? There it is. I need to give the hedge a name, my name and the date is prefilled, then I select source. I can select Upload a File or Enter Transactions One at a Time. I select Upload File, select the file type from the list, and now browse to the file location. I select it, hit Upload, and then back to Gary I think."

Gary got up from the front row, grabbed Yong and David to come along, and also asked Dexter and Ivan to come up.

"Let's talk about our plan to manage the hedge transactional data set. We know that CurrX is our ultimate system of record for the data but we need to get it there in the CurrX data format, and we need to persist every version of data given to us and each translation we make for audit purposes. I'm thinking that we uniquely identify each hedge request by user creator, customerID, and precise time stamp of initial creation, and give that a key. Then, store that data, field by field, in versioned data set on MyHedge. Once we get the data to CurrX in its standardized format, CurrX becomes the system of record for the hedge, and we only translate into it for changes, and out of it just to feed back to the customer."

Kayla, relishing her role, began to dig in. "Let's cover a few details off, shall we Gary? How do you plan to store the data? Relational or not? And what tool will do the conversions?"

Mary sat in the back of room, smiling at the interaction. It felt good to see the team getting into details and understanding the system they were about to build. It was already 2:30 PM and they were just a few steps into the first scenario, but Mary knew the simulation was already a big success.

TRANSITION TO NEXT STEPS

The first day of the simulation had ended at 5 PM as scheduled. A half-hour break for emergency email catch up had been planned before the simulation cocktail party began at Skippers, a few miles away over-looking the beach. Mary had explained to Pam, Jackson, and the other meeting planners that finishing up day one in a pleasant informal setting with a chance to informally review results, chat with partners, meet folks in person, and generally get to know each other as people was as important as the formal meeting itself.

The ideal post-meeting party needs to be set up to encourage inter-action and mixing, and Skippers had done a nice job. An open bar with two drink tickets each provided (and none more allowed – this was a business event after all), a buffet of walking-around hand-food appetizers, a few medium-sized tables with chairs and several stand-up tables, quiet music in the background and a beautiful view over the ocean to spark conversation – just right.

Pam had been worried about the cost so Mary had taken the idea to Sai, who understood, supported, and attended along with other senior executives such as Gary's boss in the IT department and the head of foreign exchange and trade finance sales.

The party had allowed reflection and connection and as day two started the team was ready to dive in again. Pam started the day by reviewing on the wall agenda where they were – just half-way through the first scenario! – and gaining agreement to spend the morning on what looked like finishing the very first scenario. The team started by discussing how many interest rate future paths they would allow Sanjeev to consider and how many hedging options for each one, and how the existing and potential future

capabilities of CurrX would constrain or limit the choices. By
noon they had worked their way
through Sanjeev being presented
a set of options, choosing one,
and his choice being registered
in MyHedge and CurrX and
a fee being taken from his company's account.

After lunch the group re-
assembled and Pam did a re-set
for the afternoon. She stood at
the wall agenda (Figure 8.3) and
began.

MEETING TIP

Have the right kind of party. The ideal
time is the end of day one of a multi-
day session. Remember that this kind
of party is still a work event – set it up
to accomplish one thing: Alignment!
The team needs time and setting for
informal dialogue, follow-up for interesting topics, and building connection.
Don't just do a dinner, and save the
celebration event for after the team
succeeds.

"We've finished just one scenario and we only have this afternoon left.
Let's look at what we have left that we need to accomplish. We had hoped
to run several scenarios with variations so that we could test boundaries of
our solution, but both the architecture review and our single
Sanjeev scenario took much longer than we had anticipated. After scenarios,
we had planned to look ahead to next steps – review the open issues and get
them assigned out and scheduled, arrange our next steps to get our plan
tightened up, and finally agree on how we communicate our results outside
of this room. We only have a few hours, should we call it a day for scenarios
and be sure we get the other items done? If we have time after that we could
start on the next scenario and see how far we get."

Mary noticed that the deliv-
ery-focused contingent was
totally with Pam, while some
of the others were more interested in making more progress
on the architectural discussion
before they got to planning
and communication. Pam,
Jackson, Gary, and Sara
wanted to get to the planning,
while David, Ivan, and Lilian
pushed to do just one more
scenario. Pam wasn't having
much luck getting consensus,

LEADERSHIP TIP

In every group it is helpful to have
a "decider" so that the team doesn't
get stuck. In practice, well-functioning
teams develop multiple deciders naturally based on topics, personalities,
and roles. It is still useful to have
a backstop who can make sure that
progress continues. Here, Mary has
made sure that Sai is present and she
can play his trump card if needed.

so Mary walked up to the front of the room to help.

"We really can't walk out of here without some forward movement towards a date," Mary said. "I talked with Sai over lunch before he had to leave, and he wanted me to thank everyone for spending the time and working together on this. But he was quite insistent that we keep our target date in mind. We need to get to market with something useful, something that declares our market intent and claims the space. I think we have a good enough idea of what we need to build to do some rough release planning now and come back together to work through the other scenarios in the coming few weeks."

To Mary's surprise and approval, the consultants from D&G – both Jackson (Agile Coach) and Melanie (Partner) had been quite restrained during the simulation. This event was not in any agile methodology or the D&G Playbook so Jackson, the D&G agile coach, had little to contribute. Now, however, Mary was talking about putting the release plan together, and Jackson knew that his methodology called for clearly specifying the Minimum Viable Product so the right selection of user stories could be slotted in sprints. He spoke up and argued that the team needed to lay out the MVP and curate the user stories for planning. Now was the time, he advocated, to do the sprint allocations he had explained at the last big team meeting (Course Correction meeting beginning on page 51).

Mary had little patience for this textbook argument although she had half expected it from Jackson. Sometimes, definition of MVP at this stage or even before was valuable as it could guide the planning. In this case, it was simply pre-mature to fix the Features leg of the Features/Cost/Time iron triangle of project management. There was too much uncertainty of what components needed to be built out, which features would be difficult versus easy, what TradeX might want to do from a product perspective, how much resources the internal CurrX team could muster for support on the desired timeframe, how quickly Gary's team would get up to speed on the new development tools. MVP, Mary knew, was a great idea but it depended heavily on understanding the capabilities of the solution technology and the capacity of the development team. Doing MVP too early was simply reverting to pre-agile upfront technology-neutral requirements specification. Deadly.

Mary did not want to just shut Jackson down so she carefully chose her words and her attitude. She agreed that the team needed to get to MVP but explained that it would make more sense to do that after the team had better information on development and integration efforts. She told a story from a past project and talked about the lesson that taught. At the end, Jackson didn't push the issue and that was the end of the D&G Playbook, at least for the day.

Pam took Mary's interventions as decisions and moved the team down to the next topic. Over the rest of the day, the team decided to hold a detailed project planning session as Mary pitched it to them and agreed that each involved party would take the next two weeks to lay out their tasks and get as much commitment from their departments as they could around the October target date. Once again Pam took overall ownership to put the planning meeting together, this time with Jackson and Kayla as devoted partners in crime. Sai came back for the last hour and the team did a quick readout to him and got his guidance on what and how to communicate status to the enterprise. Finally, Pam handed out a session evaluation sheet and asked everyone to complete it before leaving, and the day was at an end. The idea of starting a second scenario never came back up.

Mary was quite pleased with her first architecture simulation as a consultant. She had had little doubt that it would succeed; they always had in her past and this was a ripe environment for the catalyst it would provide. Mary was also excited to see Pam's emergence as a team facilitator and leader, as well as David, Judy, Sara, Gary, and Dexter all stepping up and contributing. Jackson was still trying to contribute and took Mary's guidance in public well, even eagerly volunteering to help lead the project planning work. Mary introspected a little about her own role, was it too big or too small, too directing or not directing enough. After a few thoughts, she shrugged her shoulders and retreated from her unusual introspection and reveled for a moment in the success of the day.

Signposts	• The Pacifica team holds its architecture simulation. It only gets through the first scenario and has much more to do. Nevertheless, the event is a big success driving good decisions on the system design and scope and gaining broad input and agreement on the results.
	• The next big area for the Pacifica team is to build a delivery plan.
Leadership Guides	• Get to an early milestone of broad understanding of system architecture and its interplay with business process. That sets the stage for all to follow.
	• Set up everything for rigor, alignment, and efficiency, from room configuration to activities to the cocktail party.
	• Be sure there is one or more deciders present.
Coming Up Next	We dig into a background chapter on project planning for agility before we return to see the Pacifica team plan their schedule and activities.

9

Background

Project Planning

SCRUM RELEASE PLANNING IS OFTEN NOT ENOUGH

Release planning is a standard part of scrum methodology that is often confused for agile. I certainly respect and have used basic scrum release planning to good effect; it works well in some situations. However, those situations are a bit limited. Let's do a quick review of scrum release planning and then consider the demands of projects too complex for the routine scrum methods.

Jackson, Pacifica's agile coach, described scrum release planning in an earlier chapter (beginning near Figure 6.4 in **Chapter 6**). Release planning comprises the team taking the user stories, which are assumed to be the units of development; estimating them in relative size and complexity using points; and laying them out into the sprint periods (which tend to be 2–3 weeks in length). The team discusses what order makes sense, since inevitably stories build on stories, and would usually try to do hard or uncertain tasks early in order to reduce risks.

When does this work well? There are many pre-conditions, sometimes met, sometimes not. Some conditions for success include:

- The unit of development is the user story. This is the most critical criteria, since the basis of scrum is the idea that we define, build, and test our new system one valuable piece of functionality at a time. This element is true in many cases – for a very small system entirely under the control of a single team; when building a user interface to a set of services already in place; when the various elements of the tech stack (UI, middleware, services, app servers, database) can be completely synchronized to common sprint plans around the user stories. But when a project includes software from several lightly connected teams, as in the Pacifica case including systems from multiple independent cooperating companies, the unit of development is unlikely to be a user story.

- Velocity has stabilized and is predictable. User stories are usually estimated in points, a unit of relative complexity. Teams track how many points they finish per sprint and work to adjust the planned user stories per sprint. If these relationships are stable – i.e. a team is confident in how many points they do per sprint – this point-based release planning can be accurate enough. However, this criterion is again subject to the terror of complexity. Team sizes change, and different user stories need different skills which can bottleneck, and estimates of points can be based on insufficient information or standardization.
- Teams have enough experience to be aligned on what it means to do a user story in a sprint. In loose agile, a team might start a sprint with vaguely defined and pointed user stories, flesh out the details of requirements and designs, code, and test. If the uncertainty going in results in inability to finish the work, it will carry over to the next sprint. This works well if the team is a best efforts arrangement, a fixed capacity doing work in order of importance. It doesn't work well when there is a fixed amount to be done in a fixed period for a fixed cost, which, like it or not, is the situation we find ourselves in often in business settings.
- The architecture is complete enough so that surprises are minimal.
- Dependencies outside the scrum team are minimal or well-known and planned.
- The overall environment is well-known and stable. This includes code management, test environments, test approach, migration path, etc.

In Jackson's earlier discussion of release planning, we saw Gary and Vivian, the developers, object to the simplest form of release planning. They had several concerns, including the foundational one that the user story would likely not be the unit of development, the sprint length proposed was too short to get much done, the architecture was not well enough understood, and external dependencies were not well enough accounted for. In this more complex situation, the coaching that Jackson provided was essentially turning agile into bad waterfall. The team was doing technology-neutral requirements upfront (the user stories in Jira) but not doing the detailed technical project planning that waterfall would demand. The worst of both worlds!

If a team is committed to agile values and principles and realizes that for their situation scrum release planning is manifestly inadequate, what's a leader to do? By now it should be clear: Lead! Use the principles of rigor, alignment, and efficiency, and have an extraordinarily well-prepared meeting.

LEADERSHIP OBLIGATIONS REGARDING PROJECT PLANNING

In now-routine agile (scrum) release planning as Jackson taught it, we have a good technique to identify the foundational needs (user stories organized into epics), to roughly estimate size (points), and to lay out a rough schedule based on assumptions that sometimes hold and often do not. These agile planning techniques bring some valuable rigor, alignment, and efficiency to the planning endeavor, and in many circumstances are a good fit to the planning need.

The question leaders on agile teams face then is twofold:

- Where routine scrum release planning is generally adequate, make the needed adjustments over time and help the team work within that framework effectively; and,
- Where additional planning is needed, recognize so and make it happen.

Organizational leaders often have an additional difficult row to hoe. There is a common project planning concept called the Iron Triangle. It holds that for any given level of quality, there are necessary trade-offs among scope, time, and cost. Scrum release planning, absent sufficient material, is a technique to fix quality at a high level, fix cost at scrum team size, and make scope and time the variables. It has the benefit of providing early feedback on scope and time through the user story development increment, short cycles, and sprint demos.

Those of us who have been in formal management positions know that organizations are rarely interested in the Iron Triangle. Business investment decisions are based on expected costs and benefits. Success in technology leadership includes providing reliable forecasts for time, cost, and scope for proposed initiatives, and then delivering high-quality solutions to plan. No amount of dedication to agility changes this fundamental need. As leaders it is our job to efficiently develop and share rigorous and reliable information about project outcomes, and to drive down the variability and risks of that information.

Agility is not an excuse to avoid planning.

Our rigor, alignment, and efficiency model applies well to project planning. Rigor demands that we have an excellent understanding of what we need to do, including risks and trade-offs. Alignment demands that we get as much input as needed from team members and get everyone marching to the same tune. Efficiency demands that we order the project elements in the best way (e.g. don't do big upfront requirements unconstrained by technology since we know the waste that will create; don't wait until the end to test; and many other lean/agile concepts we know well by now) and react quickly to new information.

One mechanism proven to work well to create a RAE plan is, as you might guess by now, the extraordinarily well-prepared meeting. In the remainder of this chapter and in Pacifica's imagined case study in the following chapter I will present one approach to conducting this milestone event. As I warned in the case of the architecture simulation, this is not meant to be a cookbook process definition. It is meant to illustrate one proven way to efficiently gain rigorous alignment around a defined plan. It requires that the technical architecture and associated business processes are well understood; it does not require detailed requirements (e.g. all the user stories complete in Jira, as Jackson counseled).

I've seen this kind of planning meeting conducted ranging from 25 to 125 people, in an afternoon and over a couple of days. It has been used for relatively simple projects using routine scrum planning (yes, it works very well for that, to get a level of detail and alignment deeper than the simple allocation of pointed user stories to sprints) and for highly complex projects involving multiple product vendors and an entire value stream of business entities. The basic format can be varied to fit the context over multiple dimensions: what activities to include, how broad or narrow constituencies to include, how much time to devote. The constant in all is the involvement of multiple subteams doing inter-related task planning, and timelines worked backward and forward from the meeting date to productive use.

Into the details!

ELEMENTS OF THE EXTRAORDINARILY WELL-PREPARED PROJECT PLANNING MEETING

The foundational elements of the project planning meeting are:

- **Use the nominal group technique** (see Tools Appendix, page 257). After proper set up, the group will be split up into planning teams. The teams can be split in whatever manner works best for your program. Some typical splits include by company or system; testing; change management (communication, training, procedure writing); reporting and metrics; deployment. Each group will spend time working up their milestones, tasks, and dependencies to place on a shared unifying timeline.
- **The participants and the audience.** The invitations are done with the groups in mind. Each group needs enough participants to generate and improve on ideas, at sufficient levels of authority to be committed. Five to six people per group is ideal but variation around that won't hurt. We never want more than one group to be representing one area – this isn't about generating and discussing options, it is about generating and committing to an interdependent course of action.

One major project that I led involved doing a complicated conversion of a large loan population from one system to another. Each system had many downstream interfaces and both core systems and some of the interfacing systems were provided and hosted by external vendor partners. The loan types were highly regulated. Time was of the essence, so we put together a combined architecture review, project planning, and team configuration meeting. We started by inviting about 50 people, many of whom would have to travel to attend, at a time when travel budgets were quite limited. The core team advocated with management to support the meeting and due to the importance of the project and trust in the team received an OK to proceed. As the date approached more and more people wanted to attend and we ultimately wound up with 110 people attending. We had to get a larger facility and revise the agenda to do more breakouts and read-backs. The result was by all accounts one of the best meetings many of the attendees had ever attended and created a remarkably detailed and well-understood plan, leading to a successful conversion. In retrospectives this meeting was commonly cited as a primary reason for the success. On the downside, I got chided for "Michael's enormous meeting" for a couple of years.

- **The room.** The setup here will be different than the architecture simulation setup we presented earlier (there we had a semi-circle with no tables around a large wall). We still need a large wall with chairs set up in a semi-circle around it, unencumbered by tables so that participation is maximized. However, we also need to have work tables set up one to a group. Figure 9.1 shows the ideal arrangement. Compromise as needed.
- **The timeline**. Most often at this point in a project we have a target date selected. This could have come from external demands (e.g. an irresponsible salesperson already committed something new to get a deal), from senior management seeking to make a splash (investor conference), or simply a more-or-less-reasonable aspiration. Put the timeline on the wall, going a few months beyond production release to leave room for a bug fix release or post-release events (like the celebration party). I like using actual months but if the team is

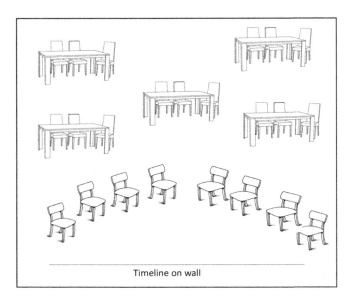

Timeline on wall

FIGURE 9.1
Room configuration for project planning.

convinced that sprint periods should be the cadence that is certainly acceptable. The timeline has rows for key project-level milestones and for each group that will be contributing their interdependent tasks. If you don't have a target date yet, just put up a timeline that will help the group figure out a date. You can use a whiteboard wall, any mostly blank wall using stickies and paper and masking tape, or the plastic roll for marking you can get at an office supply store. A sample is shown in Figure 9.2.

In this example, the team is building a web and mobile application to manage a general ledger (GL) accounting system. The web, android, and iphone teams are separate from each other, perhaps in different companies. The middleware and GL and rules management system teams are in the sponsoring organization. The meeting organizers have chosen to pre-set "code complete" and "pilot begin" milestones to guide the planning discussion. This chart, large and visible on the wall, is the framework for the entire meeting.

Notice the similarity of this chart to Jackson's sprint planning chart (Figure 6.4). Scrum release planning is a special case of this general approach.

- **The architecture.** It is imperative that going into this exercise the participants know what they need to do. Ideally the sequence is something like: identifying customer and business needs (high-level user stories/epics); architecture simulation; project planning. For complex projects it might be valuable to iterate through this a couple of times. If we aren't doing something simple enough to develop user story by user

	Mar	Apr	May	Jun	July	Aug	Sep	Oct	Nov	Dec	Jan	Feb
						Code Done	Final Test debug	Pilot		Rollout		
Web UI												
Android												
Iphone												
Middle Ware												
GL												
Rules												

FIGURE 9.2
Planning timeline for wall.

story, the team must map the user stories to the actual code modules each component team will be building.

An example here might be helpful. On a recent project the team did analysis and research to build a good backlog of user stories. Many if not most of the user stories required pieces to be built by more than one development team, and several of the key teams were in different companies with vastly different development life cycles that could not be changed for this project. Out of the architecture simulation the participants from each company went back to their home bases and worked with their governance structures to lay out the elements they would need build, the timelines they were permitted to commit to, and the processes that they would have to follow. Each team brought this information to the planning meeting. Extraordinarily well prepared, indeed!

- **Environments and code migration.** The team must have a common understanding of development and testing environments and the prospective movement of code among them, ideally diagrammed before coming into the meeting. I get accused of being a broken record on this topic, but we must remember that software development is, at least mechanically, about the movement of code from environment to environment, perfecting it as we go, until it's in productive use. The project plan needs to be focused on the code.

CONDUCTING AND CONCLUDING THE PLANNING MEETING

In the next chapter we'll see the Pacifica planning meeting, but before we dive into that story again let's touch on a few key elements in conducting and concluding the meeting.

Time and Timing

The Pacifica meeting is being done well into the project, following needs discovery and documentation and architecture planning. As such there is a fairly large team already assembled and much information on what needs to be done. The challenge is to develop the execution plan to improve chances of efficient and effective delivery. At this stage such a meeting will likely take at least half a day. It is also possible to string together several meetings such as the architecture simulation, the planning, and the team structure meetings into a day-and-a-half or two-day session earlier in the program depending on the situation. For Pacifica, a gap of a few weeks from the architecture meeting is needed for the teams to coordinate with their home bases prior to committing to delivery timeframes and resources.

Facilitating

A skilled facilitator familiar with the people and the technologies is necessary. I've seen success from project managers, the chief engineer, departmental managers, architects/engineers, and even ad hoc borrowed consultants (so long as they have time to prepare adequately). One facilitator should be the meeting lead from beginning to end, but it's nice to have other leaders take sections of the meeting.

Provide Just Enough Background and Motivation

In order to have productive dialogue the attendees need a common base of information. You might kick off the meeting by having a senior executive or a customer talk about the opportunity or problem being addressed. Then some background on the primary systems or organizations involved, a summary of prior architecture work, perhaps some reference handouts. Keep this presentation section to the minimum needed and be sure to let the real experts do the work. The sooner you can get to the activity the better.

Watch the Code

In the meeting itself, the environments used at any particular point and the migration of code among them must be prominent. The simplest case is an established team with well-regulated development and test environments and rules for migrating among them; this can be so simple that project planning of the sort we are exploring here is hardly needed. In more complicated cases there might be one or more dominant environments in which the end users will reside, connecting via middleware and services that reside in other parts of the organization or via partners or cloud-hosting. Each development team may have different sprint schedules or other development practices; one team might have several development and test environments while another might have just one or two. It matters greatly how the environments line up with each other. We'll see some of this discussed in the Pacifica example to come.

Other Roles

As in the architecture simulation, you may specify or it may evolve to have one or more roles that are unique to you. One role that was recently used in my experience is the inquisitor, which we saw Kayla play in Pacifica's architecture simulation. This role emerged completely by accident but has proven quite valuable in a couple of instances. I had encouraged a team first to do a simulation and then a planning meeting, and was fortunately able to attend both. The team was quite new to meeting techniques and to the business/technical context of the project while I was not – it was in an area in which I had experience. At the architecture simulation I was sitting near the front and as the simulation progressed, I starting asking clarifying questions. Eventually someone called me the "inquisitor in chief," in a nice way, and I checked in with the team as to whether they wanted me to keep playing that role. They did, quite surprisingly in near unison, so I happily did. When the team next did the planning meeting, they specifically asked me to play that role again. I agreed but by the time the meeting was held the team had learned so much and added sufficiently knowledgeable team members that the role proved unnecessary.

Don't be afraid to make up or recognize useful roles – this book is decidedly not a cookbook you should slavishly follow, it's an elaboration of ideas to help leaders – lead!

Outcomes

The primary outcome is a team aligned around a schedule of events and approaches to code development, code testing and migration, and production implementation including training, procedures, and communication. During the meeting the information will be developed and improved on the wall. After the meeting the project manager(s) must assemble the plan into shared useful documentation, and each responsible subteam must build out their own detail as needed. The shared plan becomes the primary management tool for the whole program, while each team delivers as they wish (or as agreed) to the shared milestones.

Enough background. Let's go see Pacifica do their project planning meeting.

Signposts	• Agile project planning often means in practice scrum release planning. This is a sufficient approach only in some less complex projects.
	• Facilitative leaders will help their teams know when scrum planning is sufficient or not; help teams adapt and continuously improve scrum; and apply more general planning techniques such as we have introduced in this chapter.
	• Organizational leaders are still expected to give accurate projections of scope, cost, and time (with quality) and deliver. A RAE approach to an extraordinarily well-prepared planning meeting is a proven technique to meet this obligation with agility.
Leadership Guides	• There is no substitute for in-person, extraordinarily well-prepared meetings to develop an effective project plan.
	• Bring together the several subgroups around a timeline from the planning meeting through productive use. Plan forward and backward.
	• Focus on the code – how does it move from development to test to production environments while being tested and debugged?
	• Agility is not an excuse to avoid planning.
Coming Up Next	The Pacifica team does its first big project planning meeting.

10

Pacifica

The Project Planning Meeting

Three weeks had passed since the architecture simulation, during which time word had spread of the successful simulation and the project planning meeting to come. As the core team prepared for the planning meeting the circle of invited participants kept growing and unsolicited requests to attend exploded. Mary had encouraged Pam to err on the side of permitting interested attendees to attend subject to a few constraints: the size of the room; fitting into the nominal groups; and providing or getting value from the session.

PREPARATION

Fortunately, the Pacifica project studio was large enough to hold the group. Pam, Vivian, Jackson, and Kayla had volunteered to put the meeting together, with Pam as overall master of ceremonies. Mary had continued to provide guidance but had tried to let the team do as much as they were able. The team had spent Monday getting the room set up and the meeting prepped, and as Mary looked around the room, it showed.

As usual, the centerpiece was the planning wall under the overhanging mezzanine. The wall was occupied by a timeline laid out in construction paper and masking tape. The timeline was on top, and rows beneath were ready for population by milestone stickies from each contributing team. The only sticky already posted in a box was the November milestone goal of first use that Sai had been insisting on since the team was spun up last Thanksgiving.

	Mar	Apr	May	Jun	Jul	Aug	Sep	Oct	Nov	Dec
Milestone										
PacificaBank.com										
Integration Bus										
MyHedge										
CurrXMaster										
Testing										
Change Management										

FIGURE 10.1
Pacifica Planning Project Wall

Mary had worked with the planning foursome to put this structure together. Jackson (Agile Coach) had argued that the timeline should show the sprints but ultimately was convinced to use the more neutral actual timeline. The various teams used different methods and release periods to deliver their software – PacificaBank.com did monthly releases, MyHedge as a new system could do whatever it chose and was being counseled by Jackson to do two-week sprints; TradeX, the CurrX vendor partner, did commercial product releases twice a year. This variability in development practices is what had led to the team being stuck a month prior and wasn't going to be resolved by synchronizing sprints and developing user story by user story.

Fifty chairs were arranged in four concentric semi-circles around the timeline. Behind the arc were six tables with eight chairs each, spaced to give as much room between them as the room would allow. Each table had a collection of meeting support paraphernalia laid out including two standing flip charts, stickies of various sizes, markers and pens, and colored masking tape. One flip chart was set up with the timeline already laid out with room for sticky tasks. Each table also had a sprinkling of chocolates, mints, and interesting little toys such as fidget spinners and yo-yos, courtesy of Pam of course.

Mary had gotten permission from Sai to put together meeting boxes for the team, and these were also on each table. Mary had taken Pam with her to shop for tool boxes or fishing tackle boxes; ultimately the tackle boxes seemed to work best. Six brightly colored plastic boxes of various sizes (the store didn't have six identical suitable boxes in stock) were filled with more supplies, including index cards, scissors, sticky dots, more tape, pens, and markers. A few of the boxes had fishing scenes on them which drew some puzzled stares and remarks.

David had been asked to put up the architecture, as developed in the simulation meeting a few weeks earlier and refined since, on the back wall for reference. His first idea was to print out the formal context diagram on a large flatbed printer. He had shown this proudly to Mary and been disappointed when she said it would not suffice. Mary had to explain that the purpose of the

LEADERSHIP TIP

Invest in meeting toolkits. They aren't expensive and encourage the right rigorous and efficient alignment activities.

architecture diagram in this context was to enable the planning participants to easily understand the pieces they needed to build and assemble so they could put dates and integrating events together. The formal notation of the context diagram was not easily consumable by non-architects, the printout though impressive was too small, and the format – a pre-printed output – was not conducive to modification in the room. Ultimately David had found his inner pre-schooler and assembled a better view out of colored paper, markers, and tape that harkened back to the simulation and that Mary had found delightful.

To the left and right of the timeline on the front wall were the props that were now becoming standard for this team: the visual agenda and an issues parking lot. On the wall next to the entrance door was a large thermometer with sticky dots and a legend sheet that was set up to gauge confidence at the end of the meeting. And, for this early morning kickoff, there was a table on the side wall set up with coffee, tea, chai, soft drinks, juice, fruit, breakfast pastries, and name tags color coded by subgroup.

Even more important than the room preparation had been the team preparation. Leaders for each subgroup had been recruited, pre-work to establish constraints and commitments for each area had been worked through, and the order of activities carefully arranged. Mary had, on several occasions, pushed the planning foursome to imagine the meeting – what could go wrong, what preparation could help it go well, what could be done in advance and what could not.

Mary spotted Pam straightening out the snacks table and wandered over to get some coffee and do a final check in while the participants filtered in. The team is doing well, she thought, and felt confident that a good day was about to begin.

MEETING KICKOFF AND MILESTONE CREATION

Pam gave her 5-minute warning just before 9:00 AM, and as the group settled into their chairs around the timeline board she went through what was now her routine way to start this kind of meeting. Because the room was going to break into subgroups, she had the attendees introduce themselves in that way. First she asked all the members of the testing team to stand up, give their name and role, and any particular concerns they wanted to address today.

From there she proceeded to the CurrX team, the integration bus, and on until everyone had a chance to introduce themselves.

Pam then turned to the wall agenda (Figure 10.2) and went through the meeting path. They would start by catching up on the architecture so everyone present, the veterans and the rookies, would understand what they were going to build. Next they would put together the project milestones, essentially the guide-rails for each subgroup's task plans. Once the milestones were agreed, the participants would split into subgroups to lay out their tasks for posting on the timeline wall. Each team would put up their tasks, one by one,

MEETING TIP

Tailor the introductions to what you want to accomplish. In this case, the meeting plan is to allow each subgroup to put their tasks together and then share and align with the others. Group identity matters much here; in other cases, less so. Name tag design is similar: color-coded tags are useful in this meeting. My personal preference is to skip personal "ice-breakers" like "what is your winter hobby?" and use work-oriented topics like "what was your favorite team and why?" if you must.

and together the group would examine and resolve dependencies and risks. "Finally" Pam explained while pointing at the large thermometer on the back wall, "each of us will express our confidence or lack thereof and we will have a real time measure of how we jointly feel about the timeline."

Planning Meeting Agenda

o   Introductions
o   Review and Understand Architecture
o   Create Milestone Timeline
o   Subgroup Task Planning
o   Add Subgroup Tasks to Timeline
o   Review and Resolve Dependencies and Risks
o   Measure Our Confidence in Timeline
o   Set Next Steps
o   Communication Agreement

FIGURE 10.2
Planning meeting visual agenda.

Pam checked for understanding and questions, took a few, and handed over to David to do a run through of the architecture (the architecture is in Figure 8.3). David asked the attendees to get out of their chairs and walk over to the architecture wall; most did, but a few did not. Pam gave the lingerers a little shove and they reluctantly joined the others at the back wall. While David was gesturing and talking, Pam checked off Introductions and Architecture on the visual agenda and freshened up her coffee.

When David wrapped up the overview, the group took a brief break and then returned to the front planning wall. Pam re-established order, showed the check marks on the first two items of the visual agenda, and introduced Gary Sunderland, the most senior Pacifica development manager and owner of MyHedge, to facilitate the next meeting segment. Gary in turn asked Janice Johanson (Pacifica CurrX Release Manager) and Lilian Kim (PacificaBank.com Analyst) to join him at the front of the room.

Mary was still in evaluation mode on Gary. He certainly was technically competent and respected by his team; her doubts were around his ability to provide the technical leadership this team needed. Mary thought Gary's perspective on the project status and path ahead had been accurate and hoped he'd do well today.

Gary was a no-nonsense, hard-working and responsible development manager, cautious and fact-based, skeptical of new things but willing to try them if they promised to work well. Mary figured he was about 45 years old with 20 years of experience, including mainframes and a variety of other platforms. Gary had been at Pacifica for a decade after moving to San Diego from Iowa when his wife, who was in the

Navy, had been transferred to the area. He was new to the public cloud but had learned enough, and played around enough, that he was enthusiastic about its potential and excited to be building MyHedge in a new way.

Gary's view on the agile studio was mixed. He found the togetherness and rah-rah attitude overdone and could not hide a smidgeon of disdain for D&G consultants in general and Jackson Maxim (Agile Coach) in particular. He had strongly embraced Mary for guiding the team to be more focused on the actual development process and for pushing to view the plan from the perspective of the software. He didn't like the lack of privacy in the studio and advocated for adding conference rooms and phone booths, which Heather and Sai were working towards. There was one overwhelmingly positive aspect of the studio, which for Gary out-weighed everything else: having direct and regular access to his business partners instead of being disintermediated by the business analysts. This in itself would result in far superior results, so if he had to put his noise-cancelling headphones on for much of the day so be it.

Gary hadn't jumped at the opportunity to lead this segment of the planning meeting when Pam and Mary had approached him. Mary wanted to see Gary in action in the technology leadership role, to see if he could succeed in it. As they talked it through, Gary had come to see the value of getting this segment right, and he overcame his reluctance to speak in front of the crowd. Mary had coached him on the segment and nervously but hopefully anticipated the performance.

"The first step in putting this plan together," Gary began, "is to get our milestones set out. At the end of the road we are going to be introducing software into production environments, so we want to focus on that software as the major milestones of the project. As we get the software milestones set, we will arrange the remaining tasks around them."

Agile Coach Jackson couldn't let this assertion of technological supremacy go unchallenged. He was generally supportive of the path Mary had them on now but hadn't let go of his designated role or the precepts of agile as he had learned them.

"Gary, you aren't saying that the technology should drive the project, are you? That would be in violation of our focus on customer needs first?"

Gary had to work to suppress a little sneer and wasn't totally successful at that. As a senior manager in large organizations Gary had learned some politic behavior and brought that out now. "Not at all, Jackson. We've already spent a few months examining customer needs and competitor offerings and have put together a technology architecture that we think can deliver. Today we want to focus on the delivery, and there really is no other way to get a good plan than to focus on the software itself. Any other approach and all we have is goals and hopes. Kind of where we've been up to now."

Gary couldn't resist getting in a little dig.

Sara Okada (Product Owner) didn't want to leave space between Gary and Jackson.

"This makes total sense to me, Jackson. Once we understand the software delivery flow, we can plan out our customer engagement tests, pilots, marketing, and training timelines as well. I think this can help us all get on the same page."

Jackson seemed satisfied with the exchange, so Gary tried to pick up where he'd left off. He walked up to the timeline on the wall (Figure 10.1, page 116) and pointed to the Milestones/November cell. He held up the sticky note he had prepared and showed the large letters saying "First Use" to the audience before placing it in the indicated cell.

"The business has set a target of first use in November of this year, and our initial sense was this was probably do-able. During the architecture simulation we developed a pretty good idea of what the first release might be and the technology pieces we need to build and integrate."

Jackson wouldn't let go. "Gary, we specified the minimum viable product in sprint zero. The user stories that we decided were needed are all coded with an MVP flag in Jira."

Gary almost suppressed a sigh of exasperation.

"That really isn't a definition of MVP, Jackson. It's a traditional waterfall approach to writing out all the requirements needed for a release without reference to the technology capabilities and constraints. We will definitely take that definition into consideration but as we build our understanding of the software elements and our joint ability to build out new capabilities, the MVP is not set in stone. Agility requires balancing desire with capability. We also need to take advantage of existing and easy-to-build valuable features to move more quickly."

This was a critical concept to Mary and she wanted to be sure the team understood it. So she got up, walked up to Gary's side, and reinforced the idea.

"I'd like us all to stop for a moment and consider what Gary just said. It's at the heart of the work we did in the architecture simulation and will be at the heart of today's planning session. True agility requires excellent understanding of customer needs and competitor offerings but that is not enough. We are building software products so the software really matters. Our success requires bringing together customer and market focus with superior software engineering and innovation.

"This is important. If we don't do both, customer focus and software engineering, we fail. Over focus on the customer and we are right back at waterfall development, with requirements in Jira as user stories instead of in bulleted lists in documents. If we over focus on the technology we risk building something that works but doesn't make us any money.

"I'm going to ask for fist to five on this. Be honest, please. If we aren't aligned on this foundational belief we need to make some changes."

Mary was pleased that the team indicated understanding; Jackson was even a five and volunteered that he didn't mean to imply that the MVP as specified was locked in stone. Comforted, Mary handed the floor back to Gary and returned to her seat.

Gary re-started the exercise. "Back to the plan. First we need to establish the milestones and get them up on the board. Once we do that, we will split into subteams to do the next level of detail.

"Let's start by planning backwards from the end, November

LEADERSHIP TIP

If it appears that the team has fundamental differences in perspective that could threaten its performance, the facilitative leader should get the issue out in the open. Flush out the naysayers and deal with their objections. My mother-in-law's marriage advice that "if it seems that something is bothering your spouse, don't ask" isn't right! Here Mary takes the bull by the horns, using her expertise and delegated decider role to leave no doubts.

first use. Do we have a date for the production release? We know that my piece, MyHedge, is net new to us and is in the cloud, so I don't have constraints on release. The integration bus is similar, we can put that into production any time we want. I think the major date constraints are on CurrX and PacificaBank.com. Let's go through those and see what we are dealing with.

"Dexter, we can start with the Pacifica implementation of the CurrX capabilities. Could you explain your production migration process and schedule?"

Dexter McDonald had been quietly lurking to the right of the planning board and now stepped under the November column with Gary.

"We are on the enterprise monthly release cycle. The November release is on Saturday the 13th. That means that our code has to be complete and in UAT by October 15," Dexter stated, while he looked at his notes. "We would have from October 15 to November 5 to do final testing and debugging and have a go/no-go decision with the rest of the monthly release on Friday November 6. That final week is what we call the quiet period where the code is frozen, and we can't make any more changes except for emergencies.

"Working back from the UAT start, the integrated testing environment will open up for the November release on September 10. We will need to get any special test data creation requirements to the testing team by around August 1 so they can pull the right production data, clean it up, and get it seeded in the right systems."

As Dexter described the dates, he took a stack of yellow sticky notes in his hand and started to peel off the cards one by one to put on the wall schedule.

"Wait a moment would you, Dexter, before you put those up," Jackson asked. "We are doing an agile project here, and I hate to say it but there is nothing agile about that schedule." Mary doubted that he hated to say that and waited for the argument to develop. Jackson was on the edge of being disruptive but was bringing up interesting points in which the team showed interest, so Mary let it continue.

"You are proposing that we do big bang of the full MVP, and we need to be code complete months before code release. And you explicitly plan for a week of doing nothing! If we are going to rapidly deliver solutions for our customers, we need to move more quickly. Can we create our own path so we can go faster? This just isn't agile!"

Mary observed the audience closely and considered intervening before Gary or Dexter choked Jackson to death. But she saw that Gary and Dexter were staying calm and let them take a shot at the objection.

Dexter was obviously put off by Jackson's passionate outburst; he had not been in the studio much and wasn't used to his well-intentioned but naïve enthusiasm. Mary could see that he was immediately put on the defensive. Nevertheless, he calmly tried to answer Jackson's objections.

"Ideally, Jackson, you are right. We would all like to be able to more quickly develop and deploy solutions. But let's look at where the issues really are – you've already spent three months on this project and are only now thinking about the software process. And the business spent years and years on other things until suddenly they have an idea they need right now! We are not the bottleneck."

Jackson started to interrupt, but Dexter wasn't nearly done.

"CurrX is integrated to more than a dozen other Pacifica and partner systems. If we make a major mistake in a release we can bring down our partners and all of our customers who depend on us. I've been in this department for 15 years, and we have fine-tuned the release cycle to get to one-month releases. I myself think we should go back to quarterly releases and just do minor configuration changes each month, but we are busting our asses to make it look like we are going faster.

"This is going to be a big change. We have decided that instead of doing custom Pacifica code, the vendor ..." he said, looking over at Ivan from TradeX, "is going to do base enhancements as part of their next release. They need to finish and test it, then we need to deal with any customization and install and test it ourselves. This schedule is already as tight as we can possibly make it."

At this point Mary concluded that the debate had gone far enough and decided to step in to take some of the heat out of the room so the day could proceed.

"Both Jackson and Dexter have good points here. Rather than further debating something that is already decided, we are going to go with the schedule that Dexter has started to put up on the board. The issue of relying on large-scale coordinated monthly enterprise releases is way beyond the scope of this team. I can assure you Pacifica's leaders are well aware of the constraints and benefits the integrated release process provides and the trade-offs and constraints it can impose on agile teams. For now, we are accepting the fact that CurrX is going to be on-release and we will adjust around that."

Mary surveyed the room, making eye contact with Sai to get a strong endorsement from the decider in the room. Sai got the message and chimed in to support Mary's direction.

"Mary's right," Sai stood up and declared. "This topic has already been discussed in the Enterprise IT Governance Committee. For now, we will not be tinkering with the monthly release process.

There is just too much to do and too much risk to make precipitous change."

Mary continued from Sai, "That's not to say that this project cannot be agile. The scrum method that many of you have learned is not the same as agile. Agile is the series of principles and values you saw in the early agile studio training. The very studio we are now in, and this meeting, helps us implement several principles such as partnerships over contracts, in-person meeting is best way to communicate, and so on. We don't have to have production releases every two weeks to be agile.

"Dexter, can you put your dates up on the wall now? And Jackson, please continue helping us be agile, we appreciate you challenging how we've always done things even though we may not want to change – yet!" Mary could feel the remaining tension in the room, but it was beginning to dissipate as Dexter started to talk about each date. Tension is OK, Mary thought, so long as it doesn't get in the way too much. It's good to debate and challenge with respect and genuine curiosity.

Dexter proceeded to put his dates up on the wall. When he was done, Gary had Lilian from the dot com team put her dates up as well. Dot com was on release for some of its areas, while new areas could be put into production more flexibly. Lilian suggested that the new screens for MyHedge go into hidden production before the November release so as new functions they could be tested in production and be ready to go on November 12 after the CurrX release. When Lilian was done, Gary had Vivian put up the key dates proposed for MyHedge, followed by the integration bus team.

The milestones were now planned backwards from first use and forward from the current date. March would focus on design, the data flows, the CurrX base enhancement, and the wireframes. In April development would be moving quickly and more data flow work would be undertaken. By May the team would have all the systems roughed in and connected in the test environment. In June and July the Pacifica team would be finishing up development, testing, and refining their pieces while TradeX finished up the CurrX enhancement. In August the CurrX base code would be available and installed for integration to the Pacifica development environment. In September and October testing and debugging would be completed and the system then readied for the November enterprise release. By Thanksgiving the first user would be uploading some hedge sets in production.

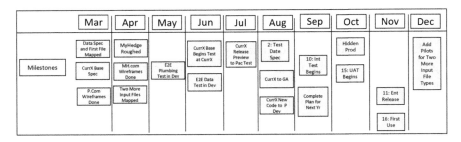

FIGURE 10.3
Partial view of Pacifica milestone plan.

As the team broke for lunch, Mary was pleased with what they had accomplished: a true, code-based milestone structure ready for the teams to fill in details in the coming afternoon session.

SUBGROUP PLANNING

Pam welcomed the team back from lunch and regained control of the room. As she now usually began meetings, she started the afternoon session by referring to the visual agenda on the front wall (see Figure 10.4).

"We're here," Pam started, pointing to the first unchecked item on the list, Subgroup Task Planning. "I'm going to ask Gary and Sara, our delivery and product managers, to walk us through the milestone timeline we just created and be sure we are all ready for our subgroups. Gary, Sara?"

Gary and Sara took over and walked through the timeline (Figure 10.3). They described each milestone briefly and took a few questions

Planning Meeting Agenda

✓ Introductions
✓ Review and Understand Architecture
✓ Create Milestone Timeline
o Subgroup Task Planning
o Add Subgroup Tasks to Timeline
o Review and Resolve Dependencies and Risks
o Measure Our Confidence in Timeline
o Set Next Steps
o Communication Agreement

FIGURE 10.4
Planning meeting afternoon agenda.

and comments. Satisfied that the participants were well grounded, Gary and Sara turned the spotlight back to Pam. Pam pointed again to the visual agenda and said, "We are now ready to start the subgroup planning work. Jackson has volunteered to help us through this segment and the milestone integration segments next. Jackson, can you set up the subgroup exercise?"

LEADERSHIP TIP

Have the team do the work. Here Pam has Gary, Sara, and then Jackson do the meeting leadership. Getting Jackson, who is only marginally aligned prior to the meeting, to lead a key section is risky but can have high payback.

When Pam had reviewed her meeting plan with Mary, Mary had initially been concerned to have Jackson lead this important part of the meeting. But Pam had convinced her that Jackson could be a help, and the subgroup planning and timeline alignment weren't all that different from standard scrum release planning. Mary had sat down with Jackson and planned out the exercise, including developing tool advertisements (explained in Appendix 2 on page 272) and coaching on how to get the rigor and alignment around the timeline the exercise sought. Mary was now hopeful that Jackson would get this done well, and anxious to see him succeed.

Jackson sprinted up to the front of the room and began to explain the exercise.

"We will split into six teams. Your name tags are color coded. Find your color table. And before someone asks, you are free to switch groups if you like. Each team has a designated reporter we arranged in advance. You will have 2 hours to prepare your detailed tasks and milestones. At 3:00 PM sharp, we will re-assemble and put together the master timeline."

Jackson now held up a sheet of paper, and Mary recognized the beginning of the Tool Advertisement she had coached. "Here is a sample of the task sheets you will complete," he explained, pointing to pre-printed forms on each table. "You each have blanks on your tables in your colors.

"This is a sample I've pre-filled as a demonstration," Jackson said as he taped a flipchart sheet surrounded by a green border onto the wall (Figure 10.5). "This is a sample I've filled out – you can see the task name here, Wireframes Complete. Below the task name a description,

```
Task:  Wireframes Complete

Description:  Wireframes for dot com and MyHedge
designed, reviewed, and ready for build

Responsible: Lilian (dot com) & Judy (UX)

Issues:  Roles of dot com vs. MyHedge designers; extend
dot com tool or use new cloud tooling
```

FIGURE 10.5
Planning task sheet sample.

the person responsible, and room for an optional list of key open issues or concerns.

"Notice how in the description I've been specific on task scope and on what task completion means. I've added who is responsible, trying to keep that limited to someone regularly involved on our team if possible. Finally, I've listed the open issues as I know them, which in this case go to how we do a unified view of customer experience design across the two different platforms.

"Any questions on how to do your task sheets? These are going to be the foundation of our plan so take your time and get the issues out on the table." This wasn't rocket science so Jackson didn't expect any confusion and he didn't get any. Mary knew that the quality of this result would depend on the diligence of the table managers they had trained for this exercise and Jackson's and her own supervision by walking around so she didn't belabor this demonstration with more exhortation to completeness.

"We are looking for something in the neighborhood of 10 to 15 of the most important tasks or milestones but have not set hard and fast rules. I'm not on a team, and neither is Mary; we will be roaming around to observe and answer questions. Feel free to engage with other tables if you need to figure out dependencies.

"Everyone ready?" No responses, so Jackson took that reasonably as assent. "Time starts, now!"

Everyone stood up from the semi-circle and headed towards assigned tables. Within a few minutes the tables were being organized by the volunteer reporters. Mary decided to start off with the CurrX team which was being led by Janice Johanssen (JJ). JJ was the CurrX release manager at Pacifica, responsible for overall release planning from request intake through requirements, design, development, testing, and ultimate inclusion of features in releases. Others at the table included Dexter McDonald (Pacifica CurrX tech lead), Ivan Ponomarenko (CurrX product manager from TradeX), Marta Smythe (Dexter and JJ's manager), and several testers, business and systems analysts, and operations partners from Pacifica and TradeX.

JJ kicked things off. She stood up at a flip chart she had prepared during lunch. It showed the CurrX milestones already on the wall – the vendor base specification complete in March, code complete early June and into testing, pre-release to Pacifica near the end of July, general availability release in August and upgrade to the Pacifica test environment with GA code, September and October testing at Pacifica, and release at Pacifica in November. Her plan was to go through each step and identify two to three predecessor or successor tasks and flesh out the issues list on each. She had planned backwards from the end to get the milestones, now she would plan forward from the current date to fill in details and reduce risks.

"Let's start with what we need to do to get the base specification complete. First, what do we mean by the spec? Is it requirements, external design, full engineering design? Ivan, can you explain what your usual process is for CurrX product development?"

Ivan proceeded to explain the CurrX product development process. For this project, Ivan explained, they expected to be developing what they called an optional feature package rather than a full-on new release. This would allow installation of new functionality without changing the base system their customers already had running, and reduce regression testing time and effort. The biggest element would be new data structures and calculation drivers to handle a larger volume of hedged data sets with fewer transaction elements, and to make the evaluation of projected hedge results re-calculate less frequently than the current base CurrX. The current CurrX was designed to support a modest number of hedge data sets with a large volume of transactions with near-constant evaluation

LEADERSHIP TIP

Each party to a program (TradeX in this example) usually has their own way of doing things, likely developed over years to meet their specific needs. Don't change if you don't have to! But do understand and ensure risks and interconnections are evident.

and visibility – current users were mostly limited to very large international corporations actively managing currency risks. The new user community would be, assuming success, a large number of smaller companies with fewer and smaller transactions, and less active management.

Ivan went on to describe the code elements they anticipated, including, at Dexter's insistence, a detailed description of the installation mechanism that Pacifica would use to add the new features to Pacifica's existing system. He explained that the March milestone spec was a standard element of the CurrX development process. It followed a milestone already passed in the last week, in which the business opportunity and rough cost estimate is presented to a management committee for prioritization and commitment. The March milestone included details on all requirements and a full technical design for consideration by the CurrX product architecture committee. This committee reviewed and approved anything heading into the CurrX base product. It met on the fourth Thursday of every month.

Ivan's description was news to most of the table. Even though Pacifica had been using CurrX for many years, the Pacifica team had never been involved this closely in the product development process. Mary was enjoying seeing the agile value of "collaboration over contracts" come to life.

Dexter absorbed Ivan's description and considered how to take best advantage of the CurrX process. "Ivan, can we help with the specification and review it before you take it to the committee? It sounds like any changes after that will be difficult."

"Sure, Dexter, that's a good idea. How about if we put a new task up two weeks before, something like 'Review draft spec with Pacifica'?" JJ was writing this onto the pink task sheet and putting Ivan and Dexter's name on it.

Dexter continued, "Are you doing any development prior to the committee decision, Ivan? Or are we frozen till you get approval?"

"We've already started some of the development," Ivan said. "But we are going to need you to sign the Statement of Work by the end of next week or we'll stop. Our agreement is to provide you with this feature, but it doesn't specify whether as part of the base or not since we can't commit until the committee meets. If they don't feel it is ready to go into base, we could deliver as custom, with best efforts to add to base as quickly as we can."

JJ jumped on this and started writing up a pink sheet for Pacifica/CurrX Statement of Work Signed. "Where are we on that?" she asked.

Satisfied that the CurrX table was on track, Mary got up to take a look at the MyHedge team.

FLESHING OUT THE MILESTONE PLAN

As the weirdly loud wall clock ticked past 3:15 PM, Pam was back under the overhanging deck in front of the wall timeline demanding attention from the intensely engaged subteams.

"I like the sound of the group," she said, "but it's time to integrate our results. This is our last task of the day, and all that's standing between us and some well-deserved drinks and food." Pam walked over to the wall agenda (Figure 10.2) and checked off Subgroup Planning as complete. "We are here," she said, pointing to Add Subgroup Tasks to Timeline, "heading to here," holding up a wine glass she'd been hiding behind her back.

"Jackson is going to bring us home. All yours, Mr. Maxim."

Jackson seemed to positively bound to the front of the group and to Mary's delight dove right into a tool advertisement.

"We are going to go group by group, starting with the dot com group and running right down the rows, through the Integration Bus, MyHedge, CurrX, then testing and change management. The reporters will come up with the group's tasks, read each one in turn, put it on the chart in its place, and take any questions or comments from the audience." He surveyed the room. "Your mission is to be sure each task is well understood and all other tasks are in proper time and dependency sequence. For each task," he continued, "besides confirming the assignment and the open issues, consider whether we have proper preparation, review, and risk reduction in place as well.

"Here is an example." Out came a purple task, a color no team had been assigned. "There is only one purple sheet, that's for me. This sheet says, extend the B&G contract so Jackson stays on team," he said with a smile, and he got a laugh from the room.

"I've assigned it to Sai, who isn't here right now so he can't refuse it, and I'm putting it in July. By July we should know we are doing great work and Sai will want to be sure I'm around to finish the job! Now, you are supposed to ask clarifying questions and be sure the date is right. Anyone?"

"OK, I'll bite," Vivian chimed in. "We probably need a recommendation from the team for Sai, don't we? Since we are trying to be self-managed? Can we add a task for June?"

"Exactly right, Viv," Jackson proudly agreed. "I take a purple sheet, add the task, and stick it up here in June." Jackson pretended to put up a new sheet, and took a little bow.

"Clear enough? Yes? Then let's start. Who is doing the dot com tasks?"

Mark Flannigan, the dot com business analyst, came to the front of the room with his colored cards.

"My first card," he began, "is for April. We have our wireframes done in March, and the MyHedge wireframes are done in April. To be sure that the user experience is consistent end to end, we need to do an end-to-end user experience walkthrough in early May. I'd really love to have more than the wireframes by then so that we could bring real customers in for a test. I know we can be ready on dot com by then, but I don't know if that's reasonable for MyHedge."

Mark looked over to the MyHedge table for a response.

Sara (Product Owner) thought Mark's idea a sparkler.

"We should be able to do that, Mark. Our April date is the final approval of our wireframe including the review by legal and compliance, so we are actually done with a pretty good draft in March as you are. We might not have our final designs, and of course they can change as Gary and Ivan get the back ends done, but we'll be far enough along to do the end-to-end test."

"And I can set up the customer test," Judy piped in from the testing table. "I'll fill out a sheet for us." Which she did, and then brought it up and stuck it in the testing row, May column. As she finished taping it up, she asked Gary, the MyHedge development manager, if he could get the dot com and the MyHedge screens actually operating with somewhat realistic data for the customer test. Gary thought so, and he turned to write up a task sheet for himself.

Jackson caught the break for the next card, and turned the room's attention back to Mark. "Your next card, Mark?"

"It looks like July is the first time we have all the code in our test environment," Mark said. "Is that right? That's when we get the new

CurrX module. Gary, is MyHedge ready for end-to-end testing then, or do we have to wait until integration testing in September?"

"Much as I'd like to be ready in July, Mark, I don't think that's going to be possible. We are aiming for code complete in Dev by mid August, which would give us time to debug and make sure things are connecting up for the September integration testing. What are you thinking about?"

"We need to get the tasks up for the security connectivity. It shouldn't take long, but that team can be really busy and we need to get on their schedule. We should do it first in the Dev environment, and then carrying it along to Integration & UAT should go easily. How about we get the connectivity going between dot com and MyHedge in Dev ready by the time you are code complete? First half of August?"

"That makes sense, Mark," Gary agreed. "Do you have a card for this already?"

"I do," Mark replied. "I was hoping it would be in July, but I'll put it in the first half of August."

"Can you tell them and be sure they are committed?" Gary asked.

"I'll take that," said Lilian Kim, the dot com systems analyst.

"And we'll get it on the master plan for communication and tracking," Mary added. "Just a quick aside, as we talk through these tasks. In addition to what's on the board, Herb is keeping track and will be responsible for assembling this into documentation to help us manage the plan. It might be helpful if you each ensure that tasks and assignments are either clearly on the board or you are sure Herb has them."

Mary had observed a lack of project management – a key framework for alignment, and useful for efficiency and rigor as well – on this agile team. The standard playbook from D&G had no role for a traditional project manager or project plan, assuming that the sprint release plan, scrum meetings, and the scrum management tool would suffice. Mary had asked Heather Gilliam (Pacifica CIO) if she could get a good technical project manager and Herb was just becoming available.

"That should be helpful," Lilian and Gary agreed.

"Anything else on the dot com to MyHedge security?" Nathan checked. "We good? OK, Mark, your next card?"

The team was on rhythm now, and spent until 5:30 PM, a half hour past the agenda end time, working through dot com, Integration Bus, MyHedge, and CurrX tasks. The agenda had contemplated finishing the timeline this day but it was not to be. As the team broke to migrate next door to Maria's for drinks and appetizers, Jackson, Pam, and Mary conferred on tomorrow. They agreed that they should spend the time it would take to finish this out and that they would ask the team to extend tomorrow's completion time from noon to 3:00 PM. They knew they would lose some of the team but figured that was the best adjustment they could make now.

Over wine and margaritas and mini-tacos in Maria's back room, Mary wandered from table to table, group to group getting feedback and scheduling information from team members. She could tell from the intensity of the conversations that it had been a good day. Mary hadn't found any irredeemable conflicts for tomorrow's extended schedule so the plan looked good.

MEETING TIP

There can be wide variation in how long it takes to do this type of interactive exercise. While the facilitators should try to keep things on time, pushing too hard can result in short-circuiting needed conversation. For Architecture Simulations and Project Planning events, it's easy to underestimate time, but hard to schedule enough to be really thorough. Meeting leaders need to make real-time judgments on how to adjust – add time, not-quite-finish and follow-up offline, add another meeting later. But don't skip the evaluation and next steps items at the end!

She spotted Jackson talking with Gary and Sai and headed over to do a final check. As she squeezed through the crowd she wondered which had been more productive, the full day of work or the 90 minutes of cocktails. It was a close contest.

SET THE INTEGRATING EVENTS

It was 10:30 AM on day two of the Pacifica project planning meeting and the time had come for Mary to use her expertise to help the team. The team had spent the morning going through the tasks for testing and change management, and Jackson had led the group through the final review and resolution for interdependencies and risks. It was now Mary's turn to teach the group about integrating events and get them onto the project.

Pam kicked off the end of Review and Resolve Dependencies and Risks by bringing the visual agenda (see Figure 10.6) up to date and explaining that Mary had one more thing to add to the plan.

"All of you are familiar with scrum, I assume?" Mary knew that most of the room had gone through the D&G training so she didn't dwell on the topic. "Then as you know, in scrum the release cycle is broken up into sprints. In each sprint, the theory goes, a complete set of user stories will be defined, built, tested and debugged, ending with a sprint demo. This happens usually every two or three weeks. As you can see from the plan we've put together, we don't have a very good map to this approach. We don't have an easy way to build out one end-to-end user story at a time; instead, we have several groups building components, following their own processes and timelines, which need to come together for testing.

"It would be different if we were just, say, building something on PacificaBank.com to existing back-end services – something where we have control of the Dev and test environments and can actually build out one user story at a time and get it tested in the sprint flow.

"We could try to force this into a periodic sprint cadence but it would be difficult. It's more natural, given the various ways each of the teams work, the lack of a common development environment, and the CurrX product development cycle, for us to do a cadence that fits with our schedules. Each team has done their best to do a schedule that gets us done on target in the most efficient way. Our challenge now is to ensure that we test things just as quickly as we can, and we build a solid base and flesh it out piece by piece until we are ready to release.

"Make sense? Jackson, any comment?" Mary turned to her now increasingly aligned junior partner.

    ✓ Introductions
    ✓ Review and Understand Architecture
    ✓ Create Milestone Timeline
    ✓ Subgroup Task Planning
    ✓ Add Subgroup Tasks to Timeline
    o Review and Resolve Dependencies and Risks
    o Measure Our Confidence in Timeline
    o Set Next Steps
    o Communication Agreement

FIGURE 10.6
Project planning visual agenda day two.

"It does, Mary. I must admit that I haven't seen this particular approach to agile, but I'm convinced that it's the best approach for this program. I don't see how we could drive this entirely with the user story backlog; we need to be more code-delivery based and bring the pieces together as they come."

"We're with you too, Mary," Gary pitched in. "We had this discussion early on before we did the architecture simulation. We couldn't figure out how we would just take the user stories and build them out and get them tested in two-week periods. What exactly are you suggesting?"

"I'm suggesting that instead of building the cadence around fixed two-week sprint demos, we do more flexibly timed integrating events. The idea came originally out of Toyota's lean product development process where each team builds its parts, whether a wheel or an axel or the transmission, and then on a set day they all come together, put it together and test it. What matters is the test, not how each team got there.

"We look at the plan and set the integrating events. For each one, Jackson can map the user stories we are satisfying, and the test team can take the acceptance criteria and put together a test plan for us to execute together. Testing will plan what data we need where and will seed it so we can do end-to-end flows. The flow of integrating events becomes the main cadence of the program."

Damien Lopez, the Pacifica test manager assigned to the project, had recently finished putting his test plans up on the wall. Mary had been quiet during his section, knowing that her introduction of integrating events would cause a lot of change. Damien noticed, of course, and engaged.

"Mary, wouldn't this require a lot more work from testing? The plan we put up on the wall was focused on the testing of the CurrX release, and then on Integration and UAT testing. I imagine we would still have to do those, but your plan would add a lot more work, wouldn't it?"

"I think so, yes, Damien. But let's see first if we have a natural set of integrating events and, if we do, consider how we approach each of them."

Damien considered Mary's proposal and assented. Mary drew a new row at the bottom of the timeline (Figure 10.1, page 116) and labelled it Integrating Events. She asked the room, "Does anyone have a suggestion for the first integrating event?"

"How about when CurrX finishes its specification – we could do an in-person review of it," suggested Dexter. Dexter was clearly excited about the opportunity to participate more directly in the CurrX product development cycle.

"Sorry, Dexter, we're fine if you want to do that, but that's not an integrating event for this exercise. We need actual software to run, test, and debug. You should make that a task sheet on the CurrX row. When is the first time we'll have enough code elements to connect up and run?"

Gary had taken to this concept from the beginning, Mary knew, and he found the first event.

"In April, Mary, we will have three file types mapped, the Pacifica Bank.com screens in Dev, and MyHedge roughed out enough to take in the files. How about we start with the customer in their accounting systems ... all three of them ... and do the initial signup and upload of the data files? That would test the plumbing from user to MyHedge, a nice chunk of the user interface, and the data file conversion software in the integration layer."

"Awesome, Gary!" exclaimed David Phillips, the program architect. "The data mapping and file conversions are always trouble, it would be great to get these tested and out of the way."

Damien was now seeing the value and saw how he could help.

"Yes, that would be valuable. It's just like the work in sprints, getting the code tested before we start the next batch of user stories, just a little different. My team could get some user PCs set up with the three kinds of accounting software, build out maybe a dozen sets of transactions we could put into each of them, and then upload each through the user interface. Then we can go look in MyHedge's database and compare the transaction sets from each brand of accounting software to be sure the conversions were done right."

Mary was excited that the team was seeing the value. "During the integrating event, we can start tracking defects, and fix them in the following week or two. During the following period testing can also run additional tests and retests of bug fixes, while getting ready for integrating event two. Damien, what do you think? More work? Is it do-able?"

"Net net, it's probably not any more work than we'd have by doing more at the end, just more constant and spread out. Let's go through the rest of the schedule and then I can evaluate the staffing I'll need.

"Remember, Damien, it's not just you and your team doing the events. Everyone will pitch in to get ready on the day we run. You will have the lead of course to write up the test plans, be sure we have the acceptance criteria captured from the user stories, and get the test data set up.

Mary wanted Damien to have a realistic sense of the workload coming his way. She was confident that it was more than he had shown coming out of subgroup planning, but likely not totally overwhelming.

Mary took a piece of plain white paper and wrote in big letters "Integrating Event One: Data Map & Load" and put it in the April column. "What day should I put on this?" she asked Damien.

Damien looked at his phone calendar, and suggested April 22; near enough to the end of the month that the end-to-end plumbing could likely be in place, but not so close to some May events that might conflict. Mary wrote "April 22" under the event title and looked around the room for concerns. A few more questions, then Mary pushed on to integrating event two.

MEETING TIP
The earlier in project planning that you can set specific dates for integrating events the better. These become the anchor dates for the entire set of work. Don't be vague; you can change them later if you must.

It took just 15 minutes to drive out a good set of integrating events. They wound up spaced roughly three weeks apart all the way through to the final start of User Acceptance Testing – the end-to-end run of scenarios after code migration to a new environment, mimicking the production release. Jackson noted that the plan wound up quite similar to sprint-end demos, just a somewhat customized schedule and a more intense focus on team testing as opposed to the mostly-for-show sprint-end demos.

Mary's final check was to examine any remaining dependencies and unresolved risks. By this time the room was flagging in attention, so she let it go quickly, resolving to look more deeply after Herb had a chance to put the plan down on paper.

Happy with the team's handiwork, Mary turned the stage back to Pam to get the day wrapped up.

CHECKING RESULTS AND SETTING NEXT STEPS

Pam returned to center stage and brought the room's attention back to the visual agenda (Figure 10.4). She pointed out that we were now near the end of the agenda, entering the activity labelled Measure Our

Confidence in Timeline. Then she asked everyone to stand up and follow her to the back wall, just to the left of the door out of the room. On the wall stood a large pasteboard with an intriguing-looking thermometer on it (see Figure 10.7).

"We are going to take a 15-minute break so you can all go outside and enjoy a little of this gorgeous day before we wrap up," Pam happily informed the group. "On the way out, we want everyone to indicate their degree of confidence in the plan we've put together here. Grab your team's colored dot and stick it on the thermometer. During the break we will total up the votes and see where we are when we get back. Please be brutally honest – we want to represent how risky we think the plan is, and also to highlight the areas where we need to do some more planning work."

Mary stood back and watched the participants circle the thermometer. As they grasped their dots Mary gave some guidance on the voting options. She preferred, she said, to rely on the verbal descriptions for dot voting on project success. Others like to think in terms of probabilities of success, so the thermometer shows percentages as well. "I'm not really sure what the numerator and the denominator of the percentages precisely mean," she said, "I try to think about the probabilities as betting odds you'd be happy to take." Some team members were placing their dots while some waited and watched, and some talked with each other mostly within the component

LEADERSHIP TIP

Regardless of how open and participative we seek to make a planning exercise, the "groupthink" pressure to conform to stated completion goal can result in unrealistic or overly risky schedules. Giving people an anonymous (if they want) chance to assess and communicate judgment can facilitate alignment and rigor. This visual tool brings efficiency as well!

groups. Within a few minutes the dot voting was done and the room was empty except for Pam, Mary, and Jackson.

Mary asked if Pam and Jackson had voted.

"Not yet, Mary," Pam volunteered. "I'm not really sure what to think. I haven't done a lot of this kind of system development work so my opinion isn't worth all that much. I was just hoping that the team was optimistic about our chances – if not, these few days didn't work very well."

"How about you, Jackson?" Mary continued. "Where did you place your dot?"

FIGURE 10.7
Thermometer confidence measurement.

Jackson, unlike Pam, had little doubt that his opinion mattered. Mary thought that on balance a good characteristic for an agile coach. He was happy to explain his vote, and even pointed to it.

"First I had to decide what color, since I wasn't on any particular team. I chose to vote with MyHedge, since that seems to be the biggest part of the Pacifica development effort. I selected 'A Good Shot at It,' 60 % probability. I definitely believe that our chances have gone up from doing the architecture simulation and this planning, although I think I would have voted for a higher probability a few months ago before I understood how complicated this was going to be."

Jackson turned the question back around to Mary, and asked if she had voted, and if so for what?

"I didn't vote, Jackson," Mary replied. "I'm trying to stay in consulting/helping mode, although it's hard for me to not jump right in. But I think the teams' assessments seem about right. Let's get the summary put together and think through how we want to facilitate the final half hour or so. We need to wrap up this day on a good trajectory."

Jackson took out his tablet and started tallying up the dot votes in a table (see Figure 10.8). After a few minutes he grabbed a flip chart and

Team	Team Average	Team High Vote	Team Low Vote
PacificaBank.com	80% Reasonably Assured	100% Total Confidence	60% A Good Shot at It
Integration Bus	80% Reasonably Assured	100% Total Confidence	40% If All Goes Well
MyHedge	60% A Good Shot at It	80% Reasonably Assured	20% A Hope but Not Likely
CurrX	40% If All Goes Well	60% A Good Shot at It	20% A Hope but Not Likely
Testing	80% Reasonably Assured	80% Reasonably Assured	60% A Good Shot at It
Change Management	80% Reasonably Assured	100% Total Confidence	60% A Good Shot at It
Total	80% Reasonably Assured	100% Total Confidence	20% A Hope but Not Likely

FIGURE 10.8
Schedule confidence by team.

wrote out the results. He studied it for a moment and then decided to add in some percentages.

Mary and Pam had been watching Jackson write the results, and after he finished Mary asked if it would be OK if she facilitated this last section of the day. "We need to dig into these results," Mary said, "and I think it's probably best coming from me." Jackson and Pam were OK with this, so as the room filled back up, Mary directed the team to take their seats in the semi-circle in front. Mary had posted Jackson's summary chart in the middle of the wall, front and center, and next to it was the dot-plastered thermometer.

"Can you all see the summary chart Jackson has put together for us?" Mary began. Most heads nodded, but a few people got up and walked closer to the chart for a better view. "Can someone summarize what it says to you?"

"Sure," Sara Okada (Product Owner) spoke up. "It looks to me that we feel pretty good about the schedule with the exception of the CurrX team. Also the MyHedge team has some concerns. And we have variation on each of the teams, with some team members confident and some quite pessimistic."

"Others?" Mary probed.

"I think that's a good summary, Mary," Dexter McDonald, the Pacifica lead for CurrX, volunteered. "It's pretty clear that the two areas with most to do are CurrX and MyHedge so it makes sense to me that we see the most risks there. I think it's great that the CurrX team is at If All Goes Well – we are doing something new together, and doing a base code change faster than ever before. We will just need to make sure that all <u>does</u> go well."

Team	Team Average	Team High Vote	Team Low Vote
PacificaBank.com	80% Reasonably Assured	100% Total Confidence	60% A Good Shot at It
Integration Bus	80% Reasonably Assured	100% Total Confidence	40% If All Goes Well
MyHedge	60% A Good Shot at It	80% Reasonably Assured	20% A Hope but Not Likely
CurrX	40% If All Goes Well	60% A Good Shot at It	20% A Hope but Not Likely
Testing	80% Reasonably Assured	80% Reasonably Assured	60% A Good Shot at It
Change Management	80% Reasonably Assured	100% Total Confidence	60% A Good Shot at It
Total	80% Reasonably Assured	100% Total Confidence	20% A Hope but Not Likely

"I like that comment, Dexter," Mary elaborated. "The worst average is If All Goes Well, which is both encouraging and scary. But we have team members in various areas who have significant concerns, which we need to listen to and see if we can address the risks early on. It's up to each of us to identify and communicate the areas where we have the most concerns, and for all of us to help address those areas.

"It's been a long couple of days," Mary turned to wrap up. She now walked over to the visual agenda and checked off the Measure Our Confidence in the Timeline row, pointing to the remaining Next Steps and Communication Agreements rows. "All we have left is to be sure we are aligned on next steps and on what and how we want to communicate these results, and then we all go home. Pam, back to you."

Pam took back the room facilitation and quickly moved through the remaining topics. Pam, Jackson, and Mary had agreed to do these quickly and get the team off home, so they had prepared suggestions and Pam just briefly tested them with the room. With little objection Pam wrapped up the meeting in just another 15 minutes, and the team was done for the day.

Remaining in the now mostly empty room were Pam, Jackson, Herb, Damien, and Mary. They sat down together at a table and talked through execution of the agreed next steps. Herb had perhaps the biggest task, putting together a detailed task plan to get the technology built and deployed for each integrating event. Damien had a lot on his plate as well, with the initial integrating event not far away; he had to get the test plan and test data ready to go. Pam and Jackson worked out how they would manage the studio, which would stay on two-week sprints to cover all the elements of the project (technical and non-technical) while Herb focused on the technology deliveries. Together they would overlap management tools so the right team members saw the right plans and tasks and had the right conversations. Mary volunteered to update Sai and Heather, the two senior Pacifica leaders, with a generally positive report.

The five leaders were tired but optimistic. They had a good technical architecture, aligned with the projected business process, and a good plan to deliver the technology. They agreed that there was some work to do to organize the team and that Pam, Jackson, and Mary would organize a follow-up meeting of a smaller group to get that done. Overall, they were proud of the results and excited to communicate to their respective leaderships, and to dive into the work.

Signposts	• Mary helps the team organize a broadly inclusive two-day event including vendor partners. • The team starts by building out a milestone path. Then they split into subgroups and each do their detailed tasks which they subsequently integrate. • Finally, the group measures their confidence in the plan – Reasonably Assured, overall, with some concerns in the CurrX and MyHedge teams.
Leadership Guides	• Building a framework for the project helps everyone be rigorous and aligned. In some agile projects, backlogs, sprints and demos work well, and suffice in this one, it took exceptional leadership to slightly modify the cadence around integrating events. • Build frameworks that fit.
Coming Up Next	• We do a background chapter on team structure. Like a good project plan, the team structure provides a framework that enables rigor, alignment, and efficiency. • Following our excursion into best agile practices, we will return to see a subset of the Pacifica team construct a team structure that fits their needs.

11

Background

Agile Team Configuration

There are few frameworks that help us achieve rigor, alignment, and efficiency more powerfully than project structure.

The Agile Manifesto itself does not speak directly to team structure, other than generally encouraging self-management, people-to-people communication, and autonomy. The popular scrum methodology added many useful ideas on roles, interactions, and governance. As scrum scaled up, additional models for structure were provided in, for example, the Scaled Agile and the Nexus frameworks. Like much of the development of agile thought, these new models focused intensively on processes and tools and only to a lesser extent on people and interactions. This chapter seeks to shine some light, in accordance with the first agile value, on the people and interactions dimension of agile teams, on their own and more importantly in the context of broader organizations.

In this chapter we will touch on:

- Agile concepts and misconceptions and the need for facilitative leadership in setting team structures.
- Setting the organizational context for agile teams: drive people, people driven, or mixed?
- Setting roles on the agile team. This section is a deep dive into roles. Agile explicitly avoids detailed role specifications in favor of the self-managed team, but is this just an abdication of leadership responsibility? What roles should we enable/establish/advocate for in order to drive rigor, alignment, and efficiency?
- Scaling beyond the single scrum team. The concept of silos and bridges is explored.

- Connecting the agile team to the broader environment. Steering groups and governance, status reports, and informal management connections will be discussed.
- The Enterprise Release and its implications for agile teams.
- Agile teams owning their own structure and continuously improving it.

AGILE TEAM STRUCTURE CONCEPTS AND MISCONCEPTIONS

As we saw in the background chapter on agile's roots, the concept of an agile team was somewhat idyllic. The development team would have business people able to work with them all the time, face to face, and give them definitive direction on customer needs and priorities. The team would self-organize, not be held to rigid plans, and be sustainable (no death-marches). Architectures, requirements, and designs would emerge from the self-organizing team, and the team would reflect and adjust itself over time. Detailed contracts and documentation, often disdained by coding-oriented developers, we explicitly devalued in comparison to collaboration and working software. Management's role, to the extent considered at all, was to give the team the environment and support they need and trust them to get the job done.

To be crystal clear, I am and have been an agile advocate (especially when coupled with lean techniques) since the ideas began to be propagated. I've written two books explaining and advocating for the use of these principles, and the groups working in my organizations in that interval have adopted agile principles and values as quickly and well as they could. Far from arguing that these ideas are wrong, here I argue that they are right, with the refinement that they need to be tempered and implemented by facilitative leaders on and off the teams seeking rigor, alignment, and efficiency.

Over the 15 years or more since the Manifesto's publication, the scrum method has grown in popularity to the extent that it is now often confused for agile itself. In addition to the formal models that have developed, some beliefs/myths/convictions have also come up. Some of these include:

- There is no need for project managers. Small self-managed teams working together through scrum masters makes project managers obsolete.
- There is no need for test managers. With testing embedded on the teams, this role is obsolete as well.
- There is no need for professional testers. Developers and business people on the scrum teams will test the software as it's developed.
- There is no need for testing. The developers will create automated tests from the detailed acceptance tests in the user stories before they write the code and won't check the code until the automated tests pass.
- Governance committees and status reports are not needed. At each sprint-end the team does a demo, showing the actual code as the best measure of progress. If there is more than one scrum team, a daily "scrum of scrums" will suffice.
- We don't need projects – instead we should organize by product (the meaning of product can be subject to infinite debate) and deliver features continuously.

All of these ideas have value and have been a powerful antidote to over-structured people-agnostic waterfall processes. Taken explicitly, however, these ideas will work more-or-less well depending on the nature of the project and the organization in which they are adopted.

Here is where leadership matters – in wise organizational leadership to encourage and enable team formation sensitive to the problems and people involved, and in strong team leadership to take ownership for their own structure and evolve to meet their needs. The Agile Manifesto contemplates the team taking this ownership but not the possible existence of wise organizational leadership.

A simple example from years ago will illustrate the point. A new agile team was spun up and the agile coach taught the then-orthodox scrum team structure. It was a small team working within a single technology stack they almost completely controlled so there was little complexity to master. However, the team soon found that the Product Owner had become an unbearable bottleneck, and that the team was having difficulty coordinating with outside technical groups like change management, security, and database administration. Most of the team reported to a single manager, and together the manager and the team adjusted roles and membership. A traditional business analyst was added to the team to be present every day and help with customer needs, priority, and solution design while the Product Owner focused more externally on business case, marketing, and interaction with business line constituents. A technical project manager was added to help with outside groups. Good leadership adjusts.

SETTING THE ORGANIZATIONAL CONTEXT FOR AGILE TEAMS

Agile team structure necessarily depends on the organizational context. Are leaders free to create and adjust team structures, or are team structures dictated and standardized to the extent that the space for leadership is severely constrained?

In my second book *Tale of Two Transformations* I present two polar opposite cases of organizations transforming to greater agility. One of the organizations is driven top-down to a specified form of agile while the other encouraged and enabled teams to evolve as best fit their circumstances. The former I called the drive people approach, while the latter I called people driven. This choice, whether made upon initial intent to become agile or constantly over the course of the journey to becoming agile, is a first obligation of facilitative leaders for agility. Neither is best in all circumstances, and neither choice is permanent. But this top-level organizational context sets the stage for how effectively leaders can accomplish agility, and the extent to which people over process is enabled to operate.

In drive people models, as a minimum the initial organizational and team structures are specified top-down. One of the best known and documented examples of a large organizational transformation of this type was done at Salesforce.com, by all reports successfully. (For a good list of references see www.velocitypartners.net/blog/2014/06/24/sales force-as-an-agile-role-model/.)

Here the role of the facilitative leader regarding team structure and people interaction is in the initial design of the process and its rollout, and then with its broad change over time. At Salesforce, many people were working on a small number of big systems, in effect one very large team, and the transformation was about changing how that team worked. Understanding the topics in this chapter would be of most use to the top-down organizational and process designers, since change on individual teams within the broader company must be aligned with others. For example, the roles and interactions of scrum masters and product managers across teams must be coordinated, and random local changes could throw the broader rhythm off. There is still room for leadership on each subteam within the overall constraints, and room for leaders to influence the overall structure and processes. But here process is quite highly valued – appropriately so.

The Pacifica case study, on the other hand, could fit the model of people driven. There are no larger enterprise demands. Sai, the business leader, and Heather, the CIO, are sponsoring a move to agility in a single area. Their goal is not organizational transformation, at least not yet; it is about adopting agile values and principles and techniques in order to accelerate business goals in a single area. From Sai and Heather down to Damien the test manager, there is great freedom to lead, freedom to create team structures and interactions as best fits their needs. Here "People," once Mary starts leading, are strongly valued over "Process" as in the Manifesto's first line. I chose to use this kind of imagined case study so I could illustrate the full range of leadership potential.

A facilitative leader at Salesforce and a facilitative leader at Pacifica would have quite different organizational contexts that would enable, constrain, and influence their scope and topics for leadership.

There are trade-offs between these two models which I can illustrate with an example. A large organization was excited about becoming more agile. Their leaders had seen some remarkable results from early adopters and had been exposed to other companies adopting vigorously. This organization had quite a large variety of systems distributed over several business units with varying degrees of interaction. Like many large organizations, the sway and pull of centralization vs. decentralization was a constant concern.

In discussing how to accelerate adoption, the discussion naturally turned to standards. The company was coming from a highly standardized required waterfall methodology, audited against, baked into planning, budgeting, and prioritization. The conversation turned to how standardized should agile become? We all need to use the same tools, don't we? We should publish new position descriptions for scrum master, product manager, release train engineer, right? Eliminate the centralized testing and project management organizations? And everyone should work on co-incident sprint periods?

For some organizations and some problems, this could of course be the right answer, as it was for Salesforce. But for it to be so, consider the implications of one of the core Agile Manifesto principles: "At regular intervals, the team reflects on how to become more effective, then tunes and adjusts its behavior accordingly." If many teams in a wide variety of problem areas, technologies, with varied partners and different histories and sub-cultures are all following the same playbook and one wants to change something, how do they do that? For example, if a team decides that a two-week sprint is too short for them and they want to try three-week sprints, would every team in the organization have to change? If one team wants to add a junior project manager role, would everyone? To what degree is standardization right for you? How much scope do you want and need to give to leadership aka "People over Process"?

I tell this story simply to illustrate that the issue of team structure is not about adopting the right team structure from agile methodologies. It is the more interesting and difficult challenge of establishing the best evolving team structure for each team in our organization seeking to implement agile principles and values.

It's not a right team structure question. It's a facilitative leadership question.

SETTING ROLES ON THE AGILE TEAM

Team structure can best be understood in our facilitative leadership context as a framework to drive our goals of rigor, alignment, and efficiency. Defining roles well is a powerful lever to empower people, develop expertise, and build effective teams. Agility also demands some new roles we will introduce, most importantly the Chief Engineer (see Chapter 3, page 19). The flip side is that defining roles poorly is a sure way to handicap agility.

Facilitative leaders have several opportunities to further "People over Process" with regard to roles. Organizational leaders can help define the roles to make cultural and institutional, influencing formal HR job families and position descriptions, salary grade levels, recruiting practices, titles, and career paths. Leaders on teams can assign or recruit roles, define missions, and reinforce or undermine responsibilities and accountabilities. Everyone can work to define their own mission and build their own expertise.

Facilitative leaders understand that they need to define roles as owners of missions. This is the three bears "just right" between the undifferentiated development team member in simple scrum and the collection-of-tasks based roles that are common in some waterfall methods. It also fits the creative nature of complex systems development.

Imagine you are a department manager and a new employee has just joined your team. If your department performs standardized work, you will likely train the employee to do certain tasks in a prescribed way. In fact, you may require standardized work and inspect behavior to be sure the defined processes are followed. As a good manager, you will also encourage that new employee to be aware of improvement opportunities and provide a path for suggestions and innovation.

On the other hand, if your department does mostly creative or non-standardized work, you will likely give your employee first a mission, and then boundaries around that mission. Employees engaged in work that is hard to standardize must think for themselves, find their own ways, and partner with their team members as they do the same.

If you are a poor manager, and are a strong believer in scrum, you might just welcome the new team member and say, "Off you go. Join the team and self-govern." This is a caricature but not by much.

Defining missions for employees is a powerful management tool that emphasizes "People and Interactions over Processes and Tools." With a mission, people are empowered to think about how to accomplish their goals with fewer constraints and more creativity and innovation than simply executing an assigned task. Roles on teams can similarly be empowering – is it your role to execute tests, or to give the OK to going to production after all known issues are identified and addressed? Is your role to be a member of the development team, or something more specific and empowering like the testing role? Leadership can empower agility by providing team members with focusing missions.

I've found that a set of roles I learned very early in my career, from Microsoft, have provided a powerful leadership framework. Let's consider those mission-based roles, and then compare to the scrum roles that are evolving as agility spreads. Neither set of roles are perfect and all need to be adjusted by leadership for each unique project and organizational context.

Microsoft Solutions Framework Roles – Still Valuable Ideas

Early in my career I was drawn to the Microsoft Solutions Framework (MSF). This certainly dates me, many of my readers will have no memory of Microsoft's early dominance of the distributed computing world! The MSF grew up out of Microsoft's successful internal development methods, more a distillation of what had grown up than an external publication of a formal documented method Microsoft development teams followed. Many years later I still believe that this is one of the best ways of thinking about roles and teams, and a useful framework as we consider agile roles.

Figure 11.1 shows a summary of the roles specified. The idea is that we create a team of peers, all focused and incentivised on delivery of a successful product, with overlapping roles for team functioning and risk management. The claim, which I find quite useful for thinking about team structure, is that all software delivery teams need all of these roles to be played, whether by one person per role, combined roles on small teams, or

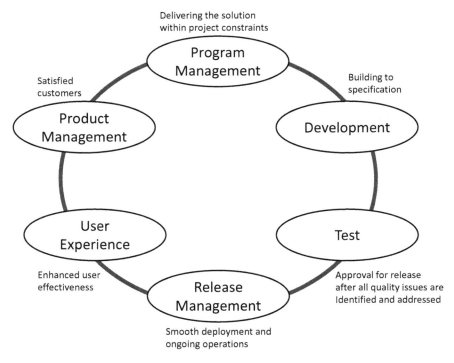

FIGURE 11.1
Microsoft solutions framework roles.

scaling out into multiple people or even teams per role. MSF also explicitly and importantly identified roles for what it called "BUMs" – business unit managers, the same role to which I refer as organizational leaders.

The team roles are:

Product Management

This role is similar to the scrum Product Owner agile role (page 156) but less overloaded with responsibilities. The primary focus and obligation of Product Management is satisfied customers. This implies that Product Management needs to understand customer needs, communicate those to the broader team, and get the product marketed/sold so it is actually satisfying customers. Unlike the scrum "Product Owner," this role is not responsible for details of the user experience or the functional design; it is an externally focused role partnering with the broader team as a peer.

User Experience

This role is entirely focused on ensuring productive use – the user interface design, ease of use, training and communication to users. Separating this from Product Management increases focus and enables different skillsets for the positions.

Program Management

Scrum mostly does away with this role, often leaving major gaps that can imperil success. The conception of this role is quite unique to MSF – the focus is "delivering the solution within the constraints." The constraints include time, cost, team size and skills, and the capabilities of the technology itself. MSF assigns the functional specification to this role, as that specification, whether a document or implicitly evolved through collaboration, is where the trade-offs are ultimately expressed. In this model business analysts are part of Program Management, rare in my experience. Program managers are also typically accountable for the project plan, status reporting/communication, and facilitating team operation (since they have functional spec and constraints). In scrum, these program management missions are designed (to give scrum credit for even thinking about this need) to be done by the team as a whole and through execution of the defined process.

Development

MSF assigns to this role the responsibility to build to the specification provided by Program Management. Build includes all aspects, including architecture – a useful reminder that most of software development is engineering, and more akin to design/build physical construction contracts than architect/build-to-design approaches. No doubt the build aspect of this role remains obvious, but "to the specification" is a matter for the team and organizational leadership to deeply consider. On some teams, especially in the waterfall model, this is true and can be appropriate. Perhaps the best example is when an onshore team does a detailed specification and design and sends offshore for coding. At Microsoft itself, this seems to have been followed more in theory than in practice. Development traditionally thought of itself as first among equals and was not a robotic coder of fully articulated specs handed to them. For example, in the early book *I Sing the Body Electronic*, about development of a children's encyclopedia at Microsoft, the developer is clearly the most important team member and brings his own perspective on functional design and even scope. He does the coding, is very smart and is far from a passive executor of others' visions (sometimes to the frustration of the other roles).

Another story illustrates enabling developers to contribute fully, not just execute the specification. This was a pure waterfall project which began with 18 months of detailed requirements work. An expensive consulting firm was hired, many workshops held, and system requirements in a technology-neutral frame were written up and handed to development. The project was replacing an existing system with the next version of the vendor's software and there were extensive conflicts between the requirements and the actual vendor technology. The vendor software did more-or-less the same things as the requirements articulate, just differently. The leadership team decided to have the developers create functional design documents translating the requirements into how they would implement the business functions in the tool they had, and then have the business analysts engage and sign off on compromise designs. This ultimately worked well, and the design documents formally superseded the requirements. If the development management had tried to build to specification, the project would most likely have doubled or more in cost and perhaps failed altogether. A good example of facilitative leadership in a totally non-agile project.

Agile principles advocate enabling developers with much more access to business people ("work together every day" "face to face") and this is one of agile's most powerful imperatives. Reducing handoffs is also a key lean concept. Scrum itself can be implemented in this way, or it can be done with the developer as robot (handed the spec as groomed user stories with acceptance criteria fully written by others, along with wireframes and full code-level details). As an agile leader, enabling creativity and efficiency of the actual people creating the software is a powerful leverage point. This has been one of the igniting elements that has led to the creation of agile studios with the entire team in one room, including the developers coding.

A recent vignette warmed my heart and illustrates what we want to create. A developer was working on the web/mobile screen that allowed a loan applicant to select the amount of the loan desired. The sprint was framed by a list of user stories, including "as a loan applicant, I want to select the loan amount and terms while seeing resulting payments so that I select the loan that fits best for me." He was at his workstation in the agile studio and showing ideas to the user experience designer and the product manager. He had roughed out examples using text input with a graphical display of resulting payment streams, and graphical input (dragging pointers) with a textual display of payment streams. The three of them worked for a few hours to get several options to a usable state. The user experience designer then assembled a group of borrowers, potential users of the system, and had them try the designs and give feedback, allowing the product manager and developer to get rigorous factual feedback to make a good design decision.

Build to specification – thankfully certainly not in these cases.

Scrum doesn't specifically draw out the Development role; it is just one of the undifferentiated team members. Lean Product Management, on the other hand, adds a Chief Engineer to lead the development function while contemplating highly differentiated roles endowed with Towering Technical Expertise. Each arrangement has potential applicability depending on context.

Test

MSF defines the role of test management in an interesting way: approval for release after all quality issues have been identified and addressed. This gives test management a sharply focused peer leadership role that broadly includes other team members. The role includes not just writing test plans and test cases, running tests, managing defects, and holding go/no-go meetings, although it typically does include all these items. The role remains even if doing full-out automated test-driven development, since it is mission-based, not task-based.

Some interpretations and even some implementations of agile techniques do not include a specified testing role. Sometimes it is subsumed in the development role, or "everyone tests." This kind of vagueness on role mission, in my mind, underserves the leadership obligation. Having a talented, experienced professional leader helping the team be rigorous, efficient, and have alignment around the milestone decision to go live is a powerful framework.

Release Management

It usually takes just one significant project for leaders to understand the value of this role, bringing focus to smooth deployment and ongoing operations. Lean thinking has, at its center, the idea of avoiding waste. One manifestation of waste in software development is our analog of inventory: untested code. The primary motivation of the foundational scrum process is that we test code immediately upon its completion thereby avoiding inventory of untested code. This method puts pressure on the release management function, building releases, moving into test environments, performance testing in line, integrating bug fixes daily or more often. Similarly if the goal is to do regular frequent releases to production, the pressure on production management escalates as well. Like the testing role, the lack of a talented, experience professional leader helping the team be rigorous, efficient, and aligned around release management would be a failure of leadership.

Scrum Roles

The simplest articulation of typical scrum roles is shown in Figure 11.2. Here we have just three roles, reflecting the Agile Manifesto principle of a self-managing team.

Product Owner

The Product Owner in scrum can be imbued with almost magical powers. In the simplest characterization, the Product Owner essentially

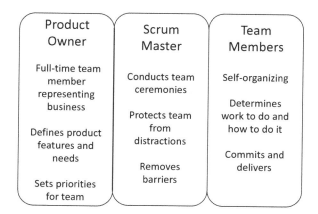

FIGURE 11.2
Typical scrum roles.

performs all of the business-side functions, while being devoted almost entirely to the team. The Scrum Alliance definition is that the Product Owner is,

> The Holder of Product Value: Determines what needs to be done and sets the priorities to deliver the highest value.

Another definition I recently saw in some agile training materials says that the Product Owner is,

> The single voice of the stakeholder community to the scrum team. This role ensures the right features are being built in the right order from the customer point of view. These team members balance competing priorities, are available to the team, and are empowered to make decisions about the product, process, or solution being developed.

From the MSF perspective, the scrum product owner includes elements of the responsibilities of product management, user experience management, and parts of program management and testing. In scrum methodology, the product owner writes all the user stories, defines acceptance criteria, prioritizes the backlog, and deals with all the business constituencies such as senior leadership, finance, sales and marketing, you name it. Amazingly this person can do all of this while spending most of their time with the development team!

My take is that this is a mythical super-human function that is not only rarely, if ever, possible but if possible, would unbalance the team and drive us right back to waterfall. Somehow the product owner can speak for all the stakeholders, simplifying the usual cacophony and complexity for the whole team so it can go code in its protected space. The roots of this concept seem to go back to the nature of the Agile Manifesto – developers defined a role that would solve the perennial problems of the business not knowing what it wants. Here, the whole development team takes on the role of "Delivery to Specification" against which I cautioned in the MSF discussion.

The product owner role obviously has important functions to be played by one or more people. However, the scrum framework is clearly problematic. The very framework undermines the idea of business and technical people working together every day (one of the Agile Manifesto principles) and substitutes the idea of a magical clarifier for the more effective team interaction of multiple roles that MSF contemplates.

Facilitative leaders must flesh out the team's framework to work well in complex and idiosyncratic organizational environments. Defining away the problem by establishing an all-powerful and all-knowing product owner could work in some cases, I suppose, but my guess is those cases are few and far between. As leaders of agile processes, we need to ensure that we don't wind up with the product management function handing off specs to the development team like we used to in the stereotypical waterfall.

Scrum Master

Like the product management role, the scrum master role has a sense of magic about it. The Scrum Alliance definition is that the scrum master is,

> The Servant Leader, protecting the Scrum process and preventing distractions.

In practice the scrum master's primary tasks are to manage the team's ceremonies – the daily scrum, which usually includes responsibility for the scrum tool administration whether whiteboards or an electronic tool such as JIRA. In the scrum meeting itself the team members are supposed to identify barriers, and the Scrum Master is tasked to remove them. The diagram in Figure 11.2 declares there is no traditional equivalent to this role, although some compare it to the traditional role of the program manager or even the organizational manager (who has no identified role whatsoever in scrum).

The software development community has taken the idea of the Scrum Master and morphed it in many ways. There are a variety of approaches to certification, ranging from a few days of simple training to much more extensive requirements. Companies have created whole job categories with scrum master 1–4 based on size of project, amount of experience, etc. Some scrum masters are mostly administrative in nature, while others are akin to the chief engineer or department manager. My preference is to have the role be flexible and defined as needed for each project, a definition arising out of the team's leadership capabilities and continually adjusted based on results.

Development Team Member

To this simple and faithful scrum model, the development team is a self-managed group that works among themselves to determine how to deliver chunks of work in frequent increments. There is no methodological barrier to defining roles however the team wishes. This is a nice idea but it fails to leverage the value of a framework entirely. What kind of roles should they define for themselves? How shall they work together to determine that? Besides following the defined scrum method, how are they to adjust how they deliver these chunks?

This is exactly the kind of situation that calls for the facilitative leader. The team needs a foundational definition of roles that could work, either coming from above in the organization or from one or more team members who have history and knowledge. Then it needs to efficiently and rigorously align around how to deliver ideas. "People over Process" demands that leaders help the team define appropriate roles so that the needed focus areas are well managed. A complex team may need development team roles of chief engineer, project manager, business analyst, system analyst, data analyst, database administrator, software engineers, test manager, test engineers, testers, user experience designer, user interface designer, compliance specialist, risk manager, financial analyst, communications specialist … the list goes on.

There are of course in the agile literature much more elaborate discussions of the various roles on teams. These can be useful to some team and team members as they work to align.

SILOS AND BRIDGES: SCALING WORK

An agile team can only be so big if it intends to fulfill some of relevant principles, such as in-person communication and working together daily. Large projects require more than one team. Let's consider a popular model for scaling as our base framework, similar to the Nexus model from Scrum.org. A simplified diagram of the basics is shown in Figure 11.3.

This model is for projects with a certain amount of unity. The amount of work is too much for one team to do effectively, so several teams are created (within the circles in Figure 11.3). The teams have some integrated work on which they work together, but each team typically has significant work it can complete on its own. The teams can be split among major features, software components, new features versus small features vs bug fixes; there are many possibilities. The separate teams are connected by Nexus Integration Team, common backlog, shared sprint cadences with common code completion and migration dates, and common ceremonies such as daily scrum and sprint reviews.

Nexus works best to deliver on single products, such as the Salesforce example highlighted earlier in the chapter. As the business context varies the model can be varied to fit. Common variations include projects or services comprising many teams but working in different technologies, different companies, some doing scrum but on different schedules with their own backlog and Product Owner, and some teams not doing scrum at all. The MyHedge project fits this more complex scaling challenge.

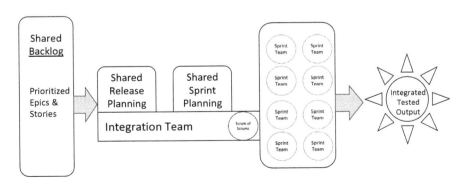

FIGURE 11.3
Scaling scrum to large projects.

I like to describe the general problem of scaling as *creating the appropriate silos with the appropriate connections (bridges)*. Helping teams get this framework right is a core obligation of organizational leadership. My simple diagram in Figure 11.4 shows the concept. The scrum master concept of protecting the team from distractions reflects a common desire of teams to be left alone to execute. The more a team can be focused on accomplishing their own mission instead of being distracted with the goals and mandates of others, the more likely they are to succeed.

On the other hand, the set of problems that a small team can accomplish on their own is quite limited. Connectors are usually required.

Scrum provides a set of connectors, well-documented in industry literature and often quite effective. These include:

- Scrum of scrums – a daily meeting of scrum masters from each team;
- Common or interlocking backlogs;
- Synchronized or coordinated sprints;
- Common development and testing environments;
- Shared tooling such as Jira for project plans, documentation, defect tracking;
- Shared retrospectives.

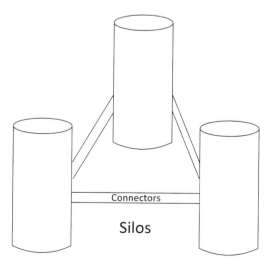

FIGURE 11.4
Silos with bridges.

Other valuable connectors within and across teams, some from lean, others from our waterfall legacy or experience, include:

- Integrating events. This is a concept from Toyota product development. The various development teams come together for a high-profile event (e.g. the car in clay); each team prepares for the event in their silo as they will, so long as they get to the event with a great contribution that fits well with the other teams. The scrum demo can be thought of as a usually small-scale integrating event. In this book, Pacifica's architecture simulation, project planning meeting, team structure meeting, and development landmark integration tests are integrating events, as are the coordinated code deliveries to the common test environment.
- Shared documentation. Documentation has gotten a bad name in the agile movement, but it remains a valuable connector if used well. Some of my continuing all-time favorites include:

  ○ The Statement of Work. There is no substitute for writing out and socializing this kind off framework when undertaking an initiative. The 1-page A3 (see Appendix 2 page 257) works well for kicking off large initiatives and for being the primary documentation for smaller ones.
  ○ The Master Test Strategy. Scrum has a built-in test strategy based on code completion in each sprint, which requires you to meet the acceptance criteria of each user story. This is a great innovation where the project is simple enough that the method just works. However, since there are wide variations from project to project, documenting the test strategy is a superior connector. This document should be prepared early in the life cycle of the team/project/ release and should cover: test stages, roles and responsibilities, tools, test data, environments, key events. It must be at the strategy level and be consumable by the broad team. This useful framework helps us efficiently and rigorously get alignment.

CONNECTING TO THE BROADER ENVIRONMENT

Both organizational leaders and leaders on project teams must concern themselves with how each initiative interacts with the larger organization.

Organizational leaders must ensure adequate engagement of themselves and other organizational leaders who influence and control resource allocations and priorities (people, money, physical space, attention). Team members must curate their relationships with their own departments (at least some will not be reporting directly to a "BUM" – business unit manager – directly owning the entire project team), and with related teams and processes on whom they depend.

Steering Groups/Governance

Scrum isn't high on these vestiges of command and control, but I remain convinced of their value if done well. For an organizational leader, well-structured governance teams can be efficient, rigorous, and get appropriate alignment on initiatives – our objectives for facilitative leadership. These groups tend to be much less obviously valuable for the teams themselves since they typically glide over topics at a superficial level and are rarely good forums to work through tough issues. Nevertheless, by keeping key organizational leaders informed, damaging unintended crosscurrents can be avoided and proper context for needed support and decisions can be maintained, which is helpful to the team. I recommend that any proposed steering groups be established formally at project inception – ideally as a section in the Statement of Work. And then be regularly evaluated for effectiveness and adjustments made as needed.

I noted that these groups aren't great at dealing with tough issues, at least most of the ones I've been aware of. The reason tends to be that there are too many people of varying organizational levels present, for one. In addition the attendees are typically not sufficiently prepped prior to a routine meeting with information and the opportunity to engage. These meetings do excel at getting initial reactions and the need for specific guidance efficiently on the table.

A small cross-functional group – not yet a team, just beginning to come together on a mission – was asked to come up with a plan to bring a new competitive capability to market. The boundaries were not entirely clear to the team. They didn't know what mattered the most: speed to market, strategic fit, low cost, fitting within the corporate planning and control rules? The team spent some time together putting together a project plan in the A3 1-page format. They got their cross-functional managers together as the steering group to review the plan. It didn't take long for the business line owner to make clear that speed to market within the strategic path was the most important need, and that she was willing to seek funding and priority outside of usual slow-moving paths. The managers of the related groups, including a few technology organizations and vendor partners, heard the direction, probed for conviction and actionability, and agreed among themselves to go forward on that basis. Details were worked out later. Here, the steering group proved efficient and conducive to alignment, while the A3 provided the framework for rigor.

Status Reporting

Status reporting is similar in nature to governance – scrum tends to focus more on in-person demos of the actual software, in line with the Agile Manifesto's wise principle that working software is the best measure of project progress. Nevertheless, well-structured and written status reports can be an efficient way of maintaining alignment with busy people outside the team – and surprisingly sometimes on the team as well.

A top organizational leader, let's call her Jane, had a team on which several people in her department were working to deliver a high-visibility solution. She was busy and had a vacation to take, so she couldn't make the time to visit the team in person for a few months. She was getting and reading status reports, however, and as the weeks dragged on, she became increasingly concerned about the project even though the status report was green and enthusiastic about story writing and grooming. Jane wanted to know if the team was on track to deliver on the date and cost for which they had been tasked but the report didn't speak directly to those issues. Her concerns peaked, Jane reached out to her people on the project and learned that there was an issue with the project manager – he was out of his depth and a personnel change was needed. By regularly exposing responsible but not deeply involved leaders to information on initiatives, it gives them the opportunity to intervene if they so choose.

Having to regularly publish a public point of view can also spark aligning incidents within a team. The same team that Jane helped by making a needed personnel change provides an illuminating example. The new project manager, let's call him Dennis, took a few weeks to get to know the team and its plans. He determined that while, from a chaining forward perspective all looked green to go, the planning for the latter stages that would get the product out the door was not far enough along. Dennis was obligated, per the terms of his defined role in Jane's department, to publish the team's point of view each week. The framework that Jane had established on status reporting facilitated Dennis' rigorous review of readiness and sparked Dennis to align with the team about his contrary observations. Ultimately the team compromised on a yellow warning status rather than Dennis' alarming red view, and work and attention were focused on the elements Dennis pointed out as wanting (and which the team as a whole agreed).

Interesting that it wasn't hard for me to come up with two examples of how a framework as simple as a required written weekly status report can support facilitative leadership.

Informal Management Connections

While there is very little in the agile literature about management at all, much less about informal relationships among managers of team members, it is hard to overstate how valuable this dimension of "People over Process" can be. Even as we seek to flatten organizations, make teams self-organizing and autonomous, and drive out hierarchy, it remains a fact for most of us that we all report to someone. We all have a boss with enormous influence over our employment, our assignments, and our pay. These bosses also have such influence over others and therefore can help us (or hurt us) in getting things done that involve more than just our own selves. Most bosses work hard to develop relationships with their peers as well, so that their influence normally goes well beyond their direct employees.

Team members should learn how to wisely use their bosses by getting and keeping them optimally involved. I won't go into this here, it's not difficult. But I will go a little deeper on boss-to-boss connection.

If you are the boss of someone on an agile team, it will often be excellent facilitative leadership for you to make an effort to establish a personal relationship with the managers of other important members of the team. The agile principles and values emphasize this: Individual and Interactions over Processes and Tools; Collaboration over Contracts; Face-to-Face Communication. Leaders are, after all, human, and good facilitative human leaders should build knowledge of and respect for each other as the foundation for success.

One more story from Jane's project, which last we saw had turned yellow due to incomplete consideration of items that needed to be accomplished a few months out. As Dennis worked with the team to address these items, one of the vendors reported that it had not fully understood what was needed and now that they did it was impossible to meet the set go-live date. Dennis knew to ask Jane for help on this rather than immediately absorb that as a fact for the team as a whole (good use of informal management-to-management communication upwards). Jane had worked hard to develop knowledge of the vendor's organization and had spent several days over a few years, and not so few bottles of wine at dinners, with senior managers there. She phoned her closest contact to get context and understanding of the internal dynamics at the vendor. What was actually getting in the way of getting the job done? Jane wound up taking Dennis and one other team member to visit the vendor's headquarters a thousand miles away to talk the issue through with the vendor's development team managers and the leader of the vendor's business line. As the issues and the stakes for both companies became clearer, the vendor chose to assign additional developers and a respected senior manager to oversee the project, and sure enough the vendor delivered as promised.

Sometimes great facilitative leadership is barely seen. But it's there – shown through "People over Process!"

Enterprise Releases

Large organizations often organize change in many interconnected systems into regularly planned period releases – monthly, quarterly, or even semi-annually. Because so many features are going into production at once, these large enterprises need to be sure that the release is not going to be disrupted by anything – they need to essentially leave no risk that failures in one project or feature are going to knock the whole enterprise back. Thus, these releases typically have long planning and prioritization cycles. For example, the final decisions for a release in June may have been done 10 months ago, August of the previous year. Release content added after last August would risk disrupting the June release by diverting tightly scheduled staff or introducing defects. Further, the limited staff for procedure documentation and training could be overwhelmed. Content added after last August stands the risk of disrupting the entire release through consuming staff in one or more areas, for example affecting content already in the release, hurrying through development and introducing serious defects or performance risks, or otherwise disrupting the orchestrated sequence of development, documentation, testing, and release to come.

To businesses wanting to be agile, the enterprise release can be a serious barrier. If your system is "on release," how do you make at least some parts of it into a silo that can decide later and deploy earlier? If your projects require changes from "on release" systems, does that mean you have to plan, architect, and design everything several quarters in advance?

We'd all like to be in simpler or more modern environments where changes can be introduced atomically, backed out easily, and resources can be allocated to our evolving needs quickly and without friction. I'd venture to say that most big organizations that rely on this kind of highly orchestrated periodic release are moving to reduce its scope and increase its flexibility. However, many organizations rely on the strong legacy of these safety-first release processes and may be living with them for some time.

An agile team doing scrum had been established to transform some internal processes to be customer facing. The first major release was planned and was moving into final testing, which made next release planning ripe. This team was doing a great job of focusing on the MVP as taught by the agile coach and leaving planning the next release for later when more information would be present. One of the options for prioritization for the next release unfortunately required material changes in an on-release system. It was February, and that system's development team was busy wrapping up the April release and fully committed to the items in the July release. The October release prioritization was just completed so the team would need to either knock out something that someone had shepherded through the demanding prioritization process or find a way to get the development to add capacity – for which there was no precedent or procedure. Else the earliest they could get what they needed was the following January, almost a full year away!

Working with the on-release dev manager, the agile team explored doing a special off-cycle release just for what they needed. If the agile team found money for contractors, could they have their own release, which would not jeopardize the broader enterprise goals? The answer turned out to be no, due to critically important constraints on code management and development and testing environments. The release process was highly tuned to be done in concert, and the risks were just too high to bear.

Organizational leaders seeking agility need to work to make the enterprise cycle more conducive to late-deciding agile teams. But in the meantime, a couple of leadership thoughts.

- It's OK, in many programs, to plan as you go. But if you are, or may be, dependent on others on a schedule you don't control, you need to be aware and plan and communicate around it. That's not falling into waterfall – it's dealing with reality.
- You may be able to get on-release development teams to serve your late-deciding agile team through cajoling, money, or friendship. But the constraints are real and remain.
- If you can see a need coming in the future, try to put the support you'll need into the on-release system, as a configuration item or in turned-off mode.

HOW SHOULD TEAMS DETERMINE AND ADAPT THEIR STRUCTURE?

We have seen in this chapter some of the options and dimensions surrounding team structure, and how important these decisions are in providing the framework for efficient rigorous aligned work. Let's wrap up this chapter by considering how a team determines and evolves its structure.

First, teams adopt existing mechanisms in their organization. There is almost always a current state, and current states usually have positive elements worth adopting. There may be commonly defined roles, centralized teams that require planned prioritization and sharing, isolated or integrated departments. Start here.

Next, team members, their managers, and stakeholders work together to define the specifics for each varying initiative. What roles are needed for this program? Who do we have to play that role? How shall we do governance? What methods should we follow? This definition takes facilitative leadership.

Finally, the team members (extended) retrospect and adjust.

To this we now turn: the Pacifica team structure, which had initially been set by the D&G consulting firm with input from Sai and Heather, is about to do a re-set based on its first few months' results' and reflection, and Mary O'Connell's expert guidance.

Signposts	• While the Pacifica team completed its project planning, we've dived into some models of team structure.
	• We've examined role definitions from early Microsoft to recent scrum and talked about some ways to scale to larger team sizes. We've worked through how an agile team can relate to the larger organization through steering groups, status reports, and informal management connections.
	• Finally, we struggled with how agile teams can deal with enterprise releases – a common process in large enterprises requiring low-risk synchronized software changes across many core systems.
Leadership Guides	• Structure matters. The common teaching of scrum (a mostly self-organizing team of formally undifferentiated roles) can certainly work but often adjustments to help with rigor, alignment, and efficiency can make a big difference.
	• Establishing mission-based roles (as we saw in the Microsoft Framework) can institutionalize improved leadership for agility. For example, the testing role is obligated to ensure that all defects are known prior to going to production; that role requires expertise in testing, leadership skills to communicate, process skills to orchestrate testing, and experience. Leaders for agility will make sure this role, and indeed all roles, is filled well.
	• Connect the team to the larger organization. Agile's admonition to leave the team alone is a nice idealistic thought and a tonic to over-control but does not optimize the enterprise's capabilities.
Coming Up Next	The Pacifica team does its project structure meeting.

12

Pacifica

The Team Configuration Meeting

ESTABLISH TECHNICAL DELIVERY LEADERSHIP

Mary had walked away from the project planning meeting with two tasks: to update Sai (business line leader) and Heather (CIO) on the results, and to work with Pam (scrum master) and Jackson (agile coach) to update the team's structure to best support the now-solidifying architecture and plan. Mary had volunteered to take the Sai/Heather task herself because she wanted their help to make a few needed changes, and she could use an update meeting as a convenient opportunity to ask.

Mary had become convinced that the team needed more technical leadership and support than they'd had so far. The consultant-led, business-focused studio had done a great job of laying the groundwork for the new business opportunity, but to get the delivery done as quickly as contemplated would take mobilization and leadership of the best that Pacifica's technology group could bring. That meant more of Heather's personal involvement was needed. But Mary didn't want to exclude Melanie Strom, the D&G partner responsible for the studio work, and risk creating a rift between the external consultants and the internal technology group. D&G were important and valuable players in the studio initiative and Mary wanted Melanie's active support and engagement. In preparation for this important conference Mary had asked Jackson to brief Melanie so any opposition could be anticipated. Jackson reported back that Melanie was happy with where the team was going, and he anticipated no obstacles.

Mary's preparation had raised the probability of success and she was pleased that the meeting with Sai, Heather, and Melanie had gone well. The three leaders knew each other mostly through larger formal settings; this was the first time they had sat down together to govern the studio work.

Heather had been involved early in the establishment of the studio and was a big supporter of agility. She had several other teams in her group doing agile

LEADERSHIP TIP

The "self-governed team" agile principle is a valuable but incomplete concept. Applying hard-earned expertise to team configuration and process and exercising the power to mobilize an organization – getting resources, priorities, money, approvals, even forgiveness – matters. Bringing this kind of external support often happens behind the scenes in senior leader-to-leader informal conversations.

in one form or another and saw the studio as a way of getting her business partners more deeply involved in solution building. Gary Sunderland (developer) was a direct report for Heather, and she had been the leader who appointed Gary to the studio and set the direction to build MyHedge on the cloud. The manager of the CurrX team reported to Heather, as did the manager of the integration technology, and the PacificaBank.com technology leader was a peer. Mary had talked with Heather twice earlier in the project to share its status and recruit talent for the studio – notably the technology project manager Herb Ratham. In short, although Heather was not a direct member of the MyHedge studio team, she had already contributed to its success and had much more to offer.

One of Mary's primary goals for the meeting had been to gain permission to position Gary as the Chief Engineer for the studio (for a refresher on this role, Chapter 3, page 19). Both Heather and Melanie were remarkably amenable, while Sai unsurprisingly had no view at all. Mary had been mildly concerned that Melanie would object that this wasn't an agile role, so the quick agreement was welcome. Heather had a few questions on how Gary's new role would differ from his already important position, which Mary addressed by explaining how all the solution delivery elements would be accountable to Gary, and Gary would be the primary spokesperson on status, needs, and risks inside and outside the team. Mary walked the three executives through some other ideas and options on team structure and previewed the team event she had in mind. Some good guidance and boundaries were provided, and a general green light to proceed was given.

Mary had followed up with a meeting with Gary and his manager Heather to be sure Gary was on board with the role. Mary explained in more detail how she had seen it work in prior endeavors and to explore any other Pacifica technical governance or organizational issues or constraints. The three of them discussed some existing tensions and constraints within the technology group, highlighting some tricky areas they would need to navigate. With good alignment from Heather and Gary on board with his stronger leadership role, Mary was ready to re-engage with Pam and Jackson and a core group together for the team structure discussion.

LEADERSHIP TIP

A technical team of peers within a larger organization can struggle to make decisions and to mobilize the surrounding organization's capabilities. Having a clear leader can help. Here, Mary ensures that clarity exists by solidifying the connection to Heather and establishing the Chief Engineer role with its formal expectations.

PREPARE FOR AND KICK OFF TEAM CONFIGURATION MEETING

After Mary's successful meeting with Heather, she gathered Pam, Gary, and Jackson into a conference room off the studio to move the team structure forward. She first explained the new role of the Chief Engineer and Heather's assignment of Gary to that role. Pam and Jackson saw the logic of having a strong technical leader able to mobilize the technology department as needed. They had both seen that the lack of strong technical leadership in the first few months of the studio had been a gap that hadn't been remedied until Mary joined. She and Gary conveyed Heather's requirements and suggestions about project team structure (such as governance team members), and again, no objections were raised. With a strong mandate from the technology team, and the already solid base from the business line through the work done to date and Sara Okada's capable direction as product owner, it was time to get the team aligned around their working mode.

Pam volunteered again to facilitate the meeting. She was a natural hostess and enjoyed being the center of attention and helping everyone to harmonize. Mary had grown to appreciate and respect Pam in this role – she knew that meeting facilitation can be done in a variety of ways, and a mostly pure process facilitator may or may not work well. In this case, Pam knew enough of the substance and had earned the trust of the participants, so she was working out well.

Together the four leaders put together an agenda, identified and invited the participants, reserved the studio and prepared the props for the session. A few days later, at 12:30 PM (Mary suggested using the afternoon so that the meeting would naturally be limited to a half-day) the group gathered in the studio and Pam assumed her usual spot at the front of the semi-circle of chairs. She quieted the room and began.

"Thanks for coming, especially our TradeX partners who flew on short notice from the Philadelphia winter. We know it's a hardship!

"As usual I've put the plan for the afternoon up on the wall," Pam started, waving towards the agenda on the wall (Figure 12.1). Mary had been teaching Pam and others on the team some of the fundamentals of meeting management and was proud to see Pam demonstrate such proficiency. "First, let's be sure we are all clear on the goal for the day, and then we can review and adjust the agenda as needed. Gary, would you mind summarizing our objectives for the afternoon?"

Gary stood up in front of the group, physically taking the larger leadership role he was now assuming.

"We've come a long way as a team, from our initial business needs-focused work in the studio and with customers, to our architecture simulation and project planning. We have a well-understood design and plan to which we are all committed. Now it's about efficient execution.

Team Structure Agenda
- o   Objectives
- o   New Team Roles
- o   Team Structure Exercise
- o   Role of the Studio / Managing the Schedule
- o   Communication of Results
- o   Next Steps

FIGURE 12.1
Team structure visual agenda.

"As we move into build, test, and deploy, we won't all be huddled in the studio every day like the core team has been these past few months. We know we have several pieces to build and integrate and test, and we will have subteams and coordination to manage. We will want to have some of us focused on specific items like testing, while others are working on marketing or procedure-writing. Each of us are responsible for this project, but each of us are also responsible to our companies or departments. Today we want to structure the team so that communication and problem-solving is as efficient as it can be, and we have the right people focused on the right items at the right times."

Gary paused to observe the room and found a rapt audience following along. He concluded his section by checking for alignment with the now well-known fist-to-five test, and then turned the podium back over to Pam.

"Pam is going to walk us through the agenda for the afternoon. Mary, Pam, Jackson, Sara, and I have been working through this for the last week, and we believe we have an approach that will work well for the afternoon."

Pam took back the floor and added to Gary's explanation for how the meeting preparation staged the afternoon for success. "There are a lot of ways for a team to organize itself," Pam explained. "As we prepared for this session, we first leveraged the work we did in the project planning workshop. Remember how we established the workstreams? That worked well so we are proposing we continue using that structure. The five of us spent some time together, and some of us also met with Sai, Heather, and Melanie for guidance and boundaries. Finally, we put this agenda and exercises together, hoping to shape today's dialogue to be efficient and productive."

"Can we get going already?" Vivian wise-cracked.

"OK, OK," Pam laughed at herself. "The first thing we are going to do is to have Mary tell us about two new roles that we will be adding to the team, and we'll see if you all want to add others as well. Then we'll get up and do an exercise to lay out the subteams, governance, and how we will all interact. If all goes well we can take a break after that, and then circle back to the role of the studio, how it will be managed, and how the schedule will be maintained – especially for those groups not in the studio. We'll wrap up by agreeing how we communicate today's results and our next steps.

"Does that meet your approval Vivian?" Pam poked.

"Yeah yeah, sure, Pam," Vivian responded, and speaking for the whole team she exclaimed, "Let's get started."

Pam walked over the visual agenda, checked off the Objectives item, and asked Mary to talk about the new roles.

INTRODUCING THE NEW TECHNICAL ROLES TO THE TEAM

"As we move more deeply into technical delivery," Mary began, "the technology team under Heather has decided to make a couple of changes in team roles. I'd like to describe them today before we get into our structuring exercise.

"The first is the addition of Herb Ratham as technical project manager. Most of you met Herb at our project planning meeting and have seen him building schedules and coordinating meetings since then. Herb's role will continue to grow as we need to manage the flow of software into environments, integration test, and so on. Herb?"

Herb stood, smiled and waved, and Mary continued.

"Technical project manager is not a standard scrum role – in fact, some agile experts even explicitly argue that the role is not needed, that it's a waterfall element. I hope you all see the value that Herb is already bringing."

Jackson had increasingly become a Mary fan and wanted to be sure there was no daylight between them. Supportively he interjected, "I for one see the value already, Mary. You're right, that role isn't part of standard scrum training, but we really need that added focus on the technical delivery schedule."

LEADERSHIP TIP

One aspect of complex teams is that members are also typically representing companies or departments. Those organizations need to have some freedom to arrange their own participation. Accept that fact, and object or adjust if needed. But give them the benefit of respect. Here Mary does not pretend the rest of the group gets a choice; this is the technology group's decision.

"Thank you, Jackson," Mary responded. "The other change is the designation of Gary Sunderland as our Chief Engineer. This is a change I recommended and have been working with Gary and Heather to bring to the team. As we stand now, we have several technical groups involved working together in and out of the studio, organized through the scrum process and their own internal collaboration. In my experience that isn't enough for a complicated program like this. Gary will be added to the formal leadership of the studio, joining Sara Okada as product owner – senior leaders. The two of them, one focused on business value and change, the other focused on product delivery, will be joined at the hip and provide formal internal drive and external representation of the team.

"Gary, could you come back up and say a few words?" Mary handed off the focus to Gary.

Gary was not a natural seeker of the limelight but could rise to the occasion when it was needed. He walked to the front left of the room and began.

"As many of you know, I've had some reservations about how we were going about the technical design and planning of MyHedge from early on in the studio work. I admit that this agile stuff is new to me, so it took a while to understand its value and what was missing. I couldn't be a stronger supporter of the studio work – I love how it brings us developers and the actual business people and customers together so we have much better information on what is needed. And since we have trainers and market and sales involved upfront, I'm expecting much better deployment to users."

Gary continued, "In our very first meeting with Mary, some of you may recall that I expressed bewilderment on how we were going to translate the good requirements into product. Jackson and I didn't see eye to eye on that – it seemed to me that the agile that Jackson was coaching expected some sort of magic to happen as user stories sprinted to demos. Mary confirmed my fears and helped us put together our really good architecture and project plan, and in the past few weeks she has been teaching Heather and me about the chief engineer role.

"Basically, it means that I'm responsible for providing leadership for our technical delivery. Kayla is going to take the direct management of the MyHedge development work so I can focus on the overall effort. We aren't changing the formal reporting relationships, but for the purpose of this program Damien, our test manager; David, our architect; Lilian, our dot com lead; Dexter, our CurrX lead; and Herb, our project manager, all report to me. I will be keeping Heather and other technology group leaders informed, and together Sara and I are the escalation points for team members. I'll also have primary accountability for the Pacifica–TradeX relationship.

"This is a new role for the bank, and I guess I've been asking for it without knowing it. As far as I can remember we've never had someone at my level dedicated to a single program – in the past that would have been seen as a demotion. So I'm asking all of you to help make this a success – I'd hate to screw up a good idea by failing at our first try."

Since the technical team had already been informed of this change, there were few comments other than Sara Okada welcoming Gary to the partnership.

Pam took back the imaginary podium, checked off New Team Roles on the visual agenda, and launched directly into the team structure discussion.

INTRODUCING THE SILOS AND BRIDGES MODEL TO THE TEAM

"Jackson, Gary and I worked with Mary to put the next exercise together," Pam started. "We decided that since Mary has facilitated this kind of team structuring meeting before, we would ask her to lead this segment. It will also give the three of us more freedom to participate, in case we have any opinions," Pam concluded, with a wry smile towards Jackson. "Mary, all yours."

Mary walked up to the center of the room and waved her hands toward the wall behind her.

"Before we start our work, I'd like to talk through the model we are going to use. I call it 'Silos and Bridges.' I'm sure you've all heard the term 'silos' used, maybe even in connection with the studio? As in, 'we need to break down the silos and get us all working together.' This is a good idea – of course we need to work together and need to break down rigid walls separating departments and companies. But we can go too far – for example, how many times in the past month have you been sitting in a meeting and thinking 'I really don't need or want to be here?' Anyone?"

David volunteered "most of them," and Kayla said "a few." The point was taken.

"I like to think about it as: we need to establish the right silos, and then connect them in efficient ways that promote rigorous and aligned decision-making."

Mary paused and directed attention to the diagram of silos and bridges she had drawn on the wall (Figure 12.2).

"Agile teaches the value of small teams, and we all know from our own experience how projects get more difficult as the number of people involved grows. There is an ideal size for each team – too small and we don't have a broad enough set of viewpoints and skills, too large and we waste each other's time or fail to communicate effectively. So, we need to build the right teams, of the right sizes and compositions. Then we want to build the bridges among the teams so each shares the information and decision-making needed by all the other teams and other constituencies that can help us succeed, such as senior management of our companies. It's a balancing act. That's what the framework on the wall is for – to help us create the right set of silos and the right set of connecting bridges."

Jackson spoke up, connecting the approach to the scrum methodology he knew and taught. "This is really just a generalization of the scrum of scrums concept, isn't it?"

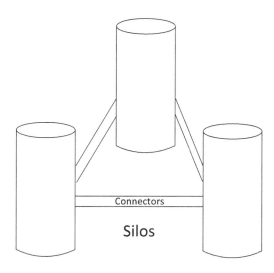

FIGURE 12.2
Silos and bridges.

"Absolutely," said Mary. "In some projects multiplying out scrum teams and collaborating via a scrum of scrums – a daily meeting of the various scrum masters – can be an effective approach. Scrum method would usually have each team be an intact development team, each regularly churning out tested user stories in a coordinated way. This works well where the broader team is able to develop by user story in a fairly atomic way, a special case most often applicable within a single large system. I believe there are plenty of cases where that isn't a great fit, so I prefer to start with a more general model. If we wind up agreeing that a scrum of scrum masters is part of our solution, so be it. Why don't we get started and see what we think? Could you all either pull your chairs closer or stand up? This will work better if we are more compact.

"Let's look at the framework I've put up on the wall for us" (Figure 12.3).

"Let's start bottom up, where we have the six teams around which we built our project plan. These are the units that are doing the bulk of the work, the ones that laid out the tasks they need to accomplish. As we agreed in our project planning meeting, these units will manage the delivery of components to our integrating events."

"Those are the silos, Mary?" Sara asked.

"Indeed, Sara. We have dot com, integration, MyHedge, the Currx team, testing, and change management. You can see how this varies from standard scrum, in that while we do have testing within each development team, we also have testing as a silo itself. Most agile coaches wouldn't like that."

"That's for sure," Jackson smiled.

"Each team will have its own roles and people filling those roles, and I'd expect the roles will vary across the teams. In order to work well together, many of us will need to know who is responsible for what on other teams, so laying this out here can help us be rigorous, aligned, and efficient. We'll go through and identify the roles and people as we start this exercise."

MEETING TIP

There is a balance between letting a team do the work themselves and helping them along. Here, the leaders have prepared a useful framework that adds rigor, alignment, and efficiency, consistent with prior work. Great leadership.

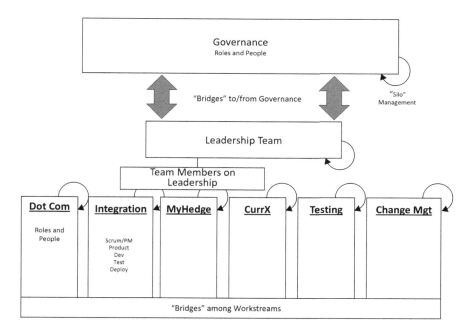

FIGURE 12.3
Team structure framework.

Sara was following along, and asked, "I suppose then we'll put the bridges in the box at the bottom?"

"Right. Once we get the silos filled up to our satisfaction, we'll go to work on the bridges. We'll identify the forums, documents, tools, and ceremonies that connect the workstreams towards our common delivery.

"Any questions so far?" Mary wanted to be sure the participants were following.

"I still have a question about the silos," said Herb. "The roles in the various teams, Mary. Do we need to be sure they align? For example, will every team have a project manager and a test manager?"

"I can see how that would be important, Herb, as you try to get us a detailed technical plan across the teams. We'll start with what roles each team typically fills, and then see if we need to modify. I would expect that just given the nature of software projects that every team would have someone doing something like project and test management, but we'll see."

Lilian from the dot com team chimed in, "We don't typically have a project manager or a test manager, Herb, because we are doing multiple scrum teams like Mary described earlier. We'll have to figure out if that works for this program or if we need to modify one or more roles to fit with the full team."

Herb wasn't satisfied. "Mary, I'd suggest we make a standard list of roles for each team to start with. If they don't fit, don't use them, but if there is something close to that using common language it could help."

Mary didn't feel strongly about this topic and would have preferred for the topic to be resolved during the exercise, but it was clear to her that the room had energy to resolve this. She asked several of the leaders their views and they seemed aligned with Herb. Sometimes the servant leader, Mary approached the model on the wall with some stickies and started pasting roles under dot com. "OK, Herb, we can do that. How about we use product manager, scrum master/project manager, development manager, test manager, deployment manager, and user experience manager? And bring your own as needed?"

"That should work, Mary," Herb agreed. Jackson, Gary, Sara, and Pam all seemed in agreement as well, so Mary moved on.

"Enough about the silos and the immediate bridges, we'll get back to these soon. I'd like to finish the overview of this tool so we can get started. If we follow upwards from the silos, we see the leadership team in the middle. These are the people who are not on any specific subgroup and are responsible for the overall project leadership or facilitation. We have Sara, as product owner; Gary, as chief engineer; Pam, as scrum master; Damien, as test manager; and Herb, as technical project manager. Often some of the leaders of the workstreams would join this group. We'll go through this once we finish with the immediate bridges."

Mary let people look around at each other and start to evaluate the soundness of Mary's suggestion. Mary thought the audience receptive.

"Next, take a look at the recursive circle on the bottom right of the leadership team and the top right of each workstream. This represents how each of these groups manage themselves – what meetings, what documents and tools, what status reports or ceremonies. Within a team there is significant freedom to customize this, especially where a team has pre-existing mechanisms. We don't want to change teams that work well; we just want to get them to work well together. Again, like the roles on a team, we need some team members to understand how some other teams work to enable efficiency and rigor."

Mary stopped for a few seconds and decided to tell a little story.

"I'd like to give you an example of the importance of knowing how other teams work. We had a vendor onsite doing an important business process. We'd work with them, give them feedback, negotiate on inputs and outputs and scope of work but never got to the point where the operation and relationship was working well. We were completely frustrated and wound up escalating to the head of the business unit at this vendor, who none of us had met prior to this scream for help. It didn't take long to learn this group's recursive circle. It turned out that this team worked by two rules which were conflicting and the resolution was killing us.

"This vendor had a recursive circle that measured unit profitability and, if negative, forced action. It had another recursive circle that enforced compliance with contracts. The company was highly de-centralized, with operating units getting attention only if one of the recursive circles sent negative signals which would often result in managers losing their jobs. It became apparent that the unit had underbid the job and was frantically doing everything it could within the terms of the contract to cut costs without tripping the specific terms of the contract. This looked to us like poor service, understaffed change resources, and continually trying to limit the scope of the work for which they were responsible. If we had understood more about the internals of their team, we might have been able to align on a fact-based plan much earlier.

"I have several other experiences where not understanding how an interacting team made decisions hurt projects, such as when the services team with which we were working depended on an aloof product management team back home. There are so many opportunities for mis-understanding. On the other hand, we don't want to all be in each other's business. It's a balancing act.

LEADERSHIP TIP

Often leaders must understand what motivates their partner teams, and how to influence. Win–win if possible, but if not, at least "we win."

"Enough story telling! We need to finish this up. The final element is governance. You can see the Governance box at the top – who in our companies needs to know our status, who can help us with resources, money, people, decisions? We'll go through and identify them, and then we need to identify the bridges to them – again, this would be meetings, demos, status reports, informal updates and such."

Mary finished up her exposition proudly. "There you have it, a general framework for team structure and interaction." Mary flipped over a chart to reveal Elements of Team Structure (see Figure 12.4). "I think we are good with the workstreams, yes?" Mary put a check mark by workstreams and pointed to the next item. "Good, then we can start with defining the roles and membership of each team."

Mary had come up with a new way to do this exercise she thought would work for this group of people at this stage. She explained, "We are a small enough group and I think this will go fast enough that we can all just do this together. We'll see if this works. Pam is passing out some small stickies and pens. I'm going to suggest that we all just start writing down roles, people, comments on internal team working, bridges, everything, and we just stick them up where they belong on the wall. It's a more chaotic approach than I'd usually suggest, but my take is that we know most of this and can get something close up on the board to review together.

"Want to give it a try?"

David Phillips (Architect) needed a clarification. "Mary, this assumes that we have the six workstreams you've identified on the wall, one central leadership group, and one primary governance group. What if we want to add a whole new team of a different structure altogether?"

"Good question, David. I'll ask the group – does anyone suggest changing the basic structure we have on the wall?" There were a few comments

Elements of Team Structure

o The Workstreams
o The Roles and People on Each Team
o The Internal Workings of Each Team (only if/as needed)
o The Bridges Among the Teams
o Members of the Leadership Team
o Format and Membership of the Governance Body
o Bridges to the Governance Body

FIGURE 12.4
Elements of team structure.

and questions, but no one had a change in mind at the moment. Mary summarized and said, "If during the exercise someone wants to add a new box entirely, let's stop and discuss it first. It's certainly not off the table if there is a good case to make."

"It's about 1:30 PM right now. How about we start with an hour and see how much we can get done? I'd suggest each person start with the team or teams

---

MEETING TIP

Setting the "right" structure for a meeting takes experience with the team and with the tools. In this case the team is familiar with each other and small enough that Mary can observe closely and assist as needed in case she needs to provide more structure. I've been doing this kind of exercise for years and am still making up new ones regularly.

---

with which they are most familiar, get those filled in, and then work on the intersections."

Everyone stood up and headed towards the wall with stickies and pens in hand. Mary noticed that several of the team members headed directly to their box (e.g. Lilian went to dot com, Kayla to MyHedge, Yong to integration, Damien to testing, Pam and Sara to change management) and quickly started pasting up small yellow and pink squares.

REVIEW AND AGREE ON SILOS AND BRIDGES

After about a half hour the initial flurry of activity turned into a steadier hum. Mary assigned one or two people to a box to de-duplicate stickies, fill in empty areas such as a few of the workstreams' recursive circles, and generally try to establish order in preparation for the full team review. As 2:30 PM approached Mary announced a break, after which the review would begin. She asked Pam, Gary, Sara, and Jackson to stay behind and put the final touches on the wall.

When the team had re-gathered, Mary drew their attention to their curated and edited handiwork on the wall (Figure 12.5).

"Congratulations, you did a fantastic job together on that. I was a little worried we'd have chaos! While you were on break the five of us," waving at the editing cohort, "did a little straightening out of the wall to make it ready for your review. Let's go box by box, see if we have any remaining

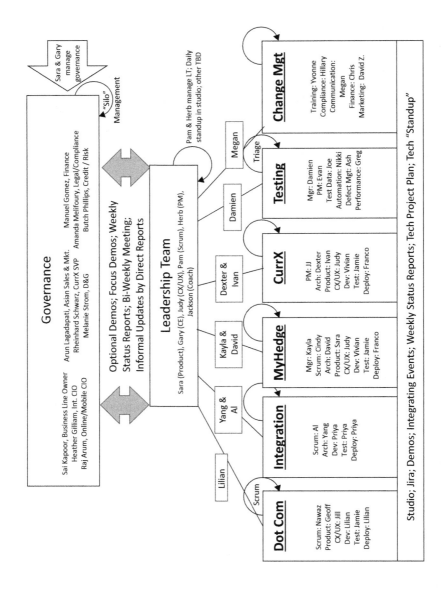

FIGURE 12.5
Proposed Pacifica project structure.

touches to add or controversies to settle, and then we'll be done with our structure for now. Remember this doesn't have to be perfect – we'll expect to make continual changes as we find we need to.

"Where would you like to start – at the top or the bottom? It really doesn't make any difference." Mary liked to give the team choices, especially when she was indifferent to the choice itself. Kind of like she did with her new baby – "would you like the creamed spinach or the creamed peas?" She tended to be less open to input when she knew the right answer.

"Could we start at the bottom with the workstreams, Mary," Herb asked. "It would really help me, as someone new, to get the full view of the larger team before we run up the ladder."

"Certainly, Herb. Everyone else OK with that?" Mary checked. Finding no resistance, she jumped right in. "Let's start with the dot com team. Lilian, can you walk us through the roles and people?"

Lilian briefly described the dot com team and how it worked. The public-facing website and mobile properties were on a two-week release cycle for routine

---

MEETING TIP

Make the team do the work.

---

changes. Every other release was reserved for larger scheduled changes, for which MyHedge would certainly qualify. Dot com ran a fairly standard scrum/Kanban practice; the team focusing on MyHedge collaboration had daily scrum meetings led by Nawaz. Geoff was product manager and would partner with Jill on the dot com team and Judy for MyHedge on the overall customer experience. Geoff would be lightly engaged on this, focused only on the introductory pages to MyHedge and how they integrated into the overall PacificaBank.com system. Lilian described how she would lead development, testing, and deployment of the MyHedge changes, and noted that she would be the only full-time dot com staff person in the studio. She would engage others as needed.

Gary and Jackson had the only questions of Lillian. Gary wanted to know how security would be handled – would dot com manage the full security review, since there was so much externallyfacing software involved, or did Gary need to engage the security group directly? Jackson wanted to talk about how Nawaz's dot com backlog and sprint schedules would line up with the studio's. Gary did need to engage directly with the security group, and he made a note to figure out where on this team structure to fit security. Without security's approval, which never seemed to come easily, this product wouldn't make it out the door. After Jackson and Lilian explored the backlog issue more deeply, it seemed that so long as Lilian managed the delivery of the needed software to the integrating events, Jackson could let Lilian keep the details of the dot com sprints within her silo.

Mary led the team through each of the workstream silos, and then moved on to the bridges below. The most animated and clarifying debate came on the last item in the bridges box, the daily tech standup. This had been added to the wall by Herb, but Jackson and Pam were definitely not onboard with the idea. Mary asked Herb to describe the item.

"Well," he started shyly, "I've only been in the studio for a week, but it seems to me that the full studio sprint management is not the right place to manage the details of the technology delivery. I'm building out the tasks for every delivery to each integrating event in a fair amount of detail. I want a daily standup where each team confirms they are on or ahead of schedule for each task, and where they can ask for help or coordinate designs or early testing. My sense is that this is too much detail for three-quarters of the studio team."

Jackson's visceral response was that this kind of separation of business and technology was exactly what Agile was meant to avoid, and he said so clearly. Pam was not quite so dogmatic about it but clearly sympathized with Jackson's take on the matter.

Mary intervened to clarify Herb's thinking and help to determine if there really was a gap here. "Herb," Mary asked, "let's go down your path a bit. Let's pretend we are in the studio and Pam is managing the overall project plan in two-week sprints. We get to sprint planning, what is in there if you are managing the technology delivery separately?"

"I don't mean that I would be doing the technology management separately, Mary. We would include in the sprint planning the items that the full team needs to know about. For example, take the first integrating event, coming up soon. In Pam's sprint plan we would certainly have the integrating event, the completion of the event test plans, the preparation for the event, and the event itself. We would cover those in each day's scrum meeting and Gary or I or another team member would give status and barriers and such. That way the interlocking of other tasks, such as the customer testing or training material preparation, can be managed in the studio."

Herb finished his argument, "But each development task for the event would be dealt with in my standup. I'd be tracking the completion of the data mapping for each file type, the coding of the transforms, the creation of the MyHedge data structures, the loading software – important items that most of the people in the studio would not be interested in. We can make the tech standup open for whomever might want to join, and I promise if most of the studio wants in we can do it in the full studio, or even, if Pam and Jackson think right, combine it right into the sprint management. But I'm pretty sure that won't happen."

Herb had been at Pacifica for several years and he felt he understood what was needed to make complicated cross-system project development work. His experience pointed to the developers and selected systems analysts and testers having their own forum. He was confident that mixing this into the full studio would hurt both elements of this project.

Gary had been listening and now exercised his new positional power as chief engineer, without being overbearing about it.

"Team, I understand the different perspectives here, so I'm going to make a decision here, and if it doesn't work we can adjust. I don't want to get the technology delivery and other program elements out of sync, but I don't want to eliminate too many 'silos.' This one I think is about building the right bridges, not eliminating all the silos. Let's try what Herb advocates."

The team accepted Gary's decision, and Mary tried to move the focus upward on the wall to the leadership and governance sections.

LEADERSHIP TIP

Knowing when to allow the team to explore and when to help the team by making a decision is part of being a leader. Gary's formal designation as chief engineer helped empower him to go beyond being a peer and bring positional power, which when sensitively exercised helps the team succeed.

She was interrupted in her intent by Damien (test manager) who apologized for the input and then suggested adding formal test plans for each integrating event as bridges. Once again Jackson initially objected.

"The tests are already in Jira," Jackson asserted, "as elaborations of the user stories. It would be duplicative and confusing to do separate test plans for each integration."

"They are different kinds of test plans," Damien contended. "The user story acceptance tests are in a sense requirements which, when we are testing user-story-by-user-story, are directly applicable as tests as well. For integrating events, we aren't testing the full system and might have parts of user stories, or collections of user stories combined in unique ways for the integration. A formal test plan for each event could identify exactly which user stories we are testing and also help us work together to prepare test environments, prepare test data, agree on degree of automated vs. manual tests, and ensure we all know our roles."

Gary stepped in again, endorsing Damien's suggestion.

"Good suggestion, Damien, let's add these to the bridges box and move on. We are already planning to do formal test plans for each integrating event, it's just a question of identifying them as bridges. Jackson, your point is taken, we don't want to duplicate work or confuse ourselves, and I'm sure Damien will welcome your input on the precise content of the integrating event test plans."

This seemed to complete the discussion of the bridges among the primary workstreams. Mary took the cue and moved the team up the wall to the Leadership and Governance teams and the bridges that connect them. Many of the attendees participated in a robust and engaging discussion.

Over the remaining hour of the meeting most of the board's content was agreed, with some good refinements:

- Kayla pointed out that the leadership team was missing an owner for production support. It wasn't clear to anyone present whose responsibility that should be, so Gary took the issue as a to do. Pam added the issue to the team's list in their tracking tool.
- There was a debate about whether the leadership team should have a separate forum from the full studio team, which already had a daily scrum meeting led by Pam. Ultimately Sara and Gary decided that they wanted to try a weekly 1-hour in-person meeting in the conference room off the studio. They both felt there would be issues that would be inappropriate for the full studio, e.g. personnel performance issues. Sara asked that Pam take responsibility to do agendas for these meetings and that she publish via email the prior evening.
- The composition of governance was also a topic of lively debate. Sara proposed adding several corporate functions that she had been dealing with one-off, which she now much preferred to bring together in one regular forum. This included finance, credit/risk, and legal/compliance. Dexter had added their vendor partner TradeX, which provoked opposition from several teams who argued that governance

should be just Pacifica. Feelings were strong on this topic, so Mary volunteered to take the issue to Sai and Heather to get their decision on it. Mary would be strongly advising to include the vendor.

- Jackson asked where Mary fit on this chart – was she governance, leadership, or what? Mary thought this an excellent question, one she had, in fact, been asking of herself. Mary described her engagement from Sai, which was to observe, evaluate, and help the team. Her current engagement was coming to an end and she needed to re-up if the team and/or Sai wanted her to continue. Now that we've come this far, she asked, what would you like of me? The consensus was that they would like her to continue in her advisory role, visiting the studio regularly, attending the leadership and governance meetings, and being kind of a super technical agile coach along with Jackson. Mary said she would be honored to stay on in that role and promised to talk it over with Sai and get back to the team. Sara wrote up a new sticky labelled super coach, put Mary's name on it, and pasted it into the Leadership Team box.

By this time, it was getting late and Pam knew it was her turn to wrap up the day. She drew the team's attention back to the visual agenda (see Figure 12.6) and checked off the completed items.

"All we have left," said Pam, "is to agree on how we want to communicate these results and what our next steps will be."

The team, now stronger than before the day began, polished off their work as the clear San Diego day came to a close.

Team Structure Agenda

✓ Objectives
✓ New Team Roles
✓ Team Structure Exercise
✓ Role of the Studio / Managing the Schedule
o Communication of Results
o Next Steps

FIGURE 12.6
Team structure meeting ending visual agenda.

Signposts	• The Pacifica team has worked together in an extraordinarily well-prepared meeting to create a structure of silos and bridges that will work well for them.
	• An appropriate governance structure, multi-level in this case with a core team and a governance group of the most senior executive is designed.
Leadership Guides	• Team and governance structure matter. Think of team structure as one of the most important frameworks that enable RAE leadership.
	• Putting together the optimal structure requires leadership. The example of how Pacifica achieved this outcome via a well-prepared and facilitated meeting and frameworks prepared in advance by leaders is one way to accomplish this.
Coming Up Next	As the Pacifica project progresses, we will more briefly examine a series of routine meetings. We'll do background of the topic followed by Pacifica demonstrations as the project progresses. First, the scrum.

Section 3

Background / Pacifica

Routine Meetings

I'm going to change up the approach in this section. Whereas in past sections topical background and the Pacifica demonstration each had their own chapter, in this section background and demonstration are combined in single chapters. I'm doing this because each section is shorter than the long treatments in prior sections. Each chapter will start with background information, and then we'll see a demonstration in the Pacifica project. The pace picks up.

For those readers who do not work in an environment of lots of people and lots of meetings, I'll repeat a statement from the meetings chapter: the background and scenes can serve as a metaphor for other types of people and leadership interaction. For example, we have a governance meeting – if instead of a meeting you just have interaction with your manager, our background information advocating how to think about your audience and determine your agenda are equally applicable whether the forum is a big meeting or a tête-à-tête with your boss.

The topics we'll cover in this section are typical of many projects and include:

- The daily scrum meeting;
- A meeting or a systems analyst?;
- Demonstrations;
- Governance meetings;
- Teleconferences;
- Leading as a participant.

We'll also see the Pacifica project proceed, get in some trouble, deal with it, and approach delivery.

Because the chapters are short, there won't be a chapter-end Signposts/Leadership Guides/Coming up Next summary until the end of this entire section.

We'll start with the daily scrum meeting.

13

The Daily Scrum

BACKGROUND

The daily scrum is probably the most commonly used agile technique. In fact I find that many people confuse scrum with agile itself. Better to think of scrum as a technique that effectively implements some of the agile principles such as "face-to-face communication is best" and "business and technology work together every day."

There is a lot of literature and training on scrum so I won't go deeply into it – just a brief refresher. First some planning is done, often called sprint zero. In sprint zero the top-level requirements are gathered, stated as user stories, some architecture work is done, and estimates are created often in an abstract concept known as points. User stories are allocated to development periods called sprints, usually two to four weeks. When a sprint starts the user stories are converted to tasks and the team commits to finish that work in the sprint. "Finish" typically has a specific meaning, most often completing detailed requirements and design, coding, testing, and debugging. Sometimes requirements and detailed design are completed in one sprint and code and test in the next. Over the course of the sprint the team allocates tasks to team members, does the work, and hopefully finishes the committed work.

The daily scrum is a stylized stand-up meeting. The meeting itself is often called a ceremony and the scrum master is the master of ceremonies. Scrum has certain rules, such as size (e.g. no more than 10 people), duration (15 minutes), and participation (only team members can speak, managers can only observe). It also has a standard agenda – the scrum master goes person by person asking three questions:

1. What did you get done yesterday?
2. What are you doing today?
3. Is anything in your way?

Some teams will integrate burn down tracking into the questions – projecting the time to complete each remaining task so the team can track progress and assess the outlook to meet the committed content completion. Other teams formally track the answers to question 3, "What's in your way?" as impediments.

As a facilitative leader your job, whether in a formal organizational or team management position or as team member, is to adjust these rules to your situation. If a senior organizational leader comes to visit do you let her speak? If the team wants to extend to a half hour and spend some time on solutioning and not just identifying barriers, is that OK?

PACIFICA SCRUM MEETING

At 9:30 AM Pam rang the ship's bell in the studio and the daily scrum came to order. Several weeks had passed since the team configuration meeting and the rhythm of the project had started to solidify. The daily full studio scrum was the primary project cadence each day, driven by the scrum board now at the center of the front wall (see Figure 13.1).

The team had agreed that the primary studio scrum board would be aimed at managing the studio itself. This meant in practice the activities of the change management team and the intersection of the technology delivery progress and the people in the studio. The details of the technology items leading up to each integration event were not being managed on the studio board, as each of the various development teams followed their own development processes. Interestingly enough, three of the key development teams – dot com, MyHedge, and integration – all used scrum methods themselves but on somewhat different schedules and approaches. For the purposes of the studio, the scrum boards of these three groups and the more traditional project plans of the CurrX team were neither here nor there – the studio saw only the stories and tasks that affected the room in general.

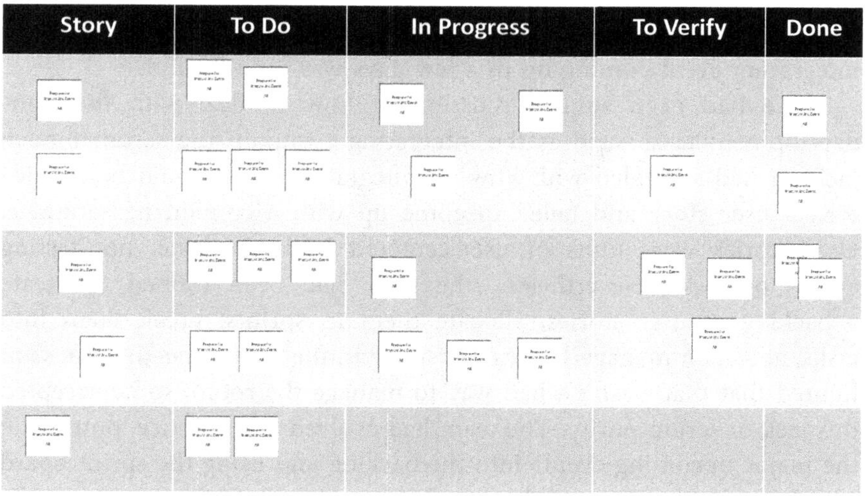

FIGURE 13.1
Pacifica scrum board.

The studio scrum board was created each week, usually mid-morning Friday, by what had become the project organizing group: Pam, Herb, Jackson, Gary, Sara, and Megan (change manager). Each week others could attend if they chose, and usually several did, most often Damien, Judy, and Mary.

Pam was surrounded by all the team members present in the studio. Today there were around 40 people, many more than standard scrum methodology would suggest, but Gary and Sara had made the decision to try it this way. This week the studio population had grown to include several additional team members from dot com, MyHedge, and integration as prep for when the first code integration event reached its height. There were also a couple of visitors, which had become normal as word of the ambitious new approach spread through Pacifica.

There were a variety of subteams and functions represented in the studio and the goal of the daily scrum was to keep everyone organized. Much additional task management and communication happened in other ways; this was the overall daily cadence setter, more communication and alignment than detailed task management and impediment removal.

Pam began as she always did, calling on one of the team members to answer the three questions. Today she started with Herb, since the integrating event coming up in a few days was top of mind.

There had been some discussion of how to deal with the non-development tasks such as the integrating events in the scrum model; Jackson had struggled with how an integrating event could be considered a user story and failed to come up with a convincing rationale. User stories were units of user-centered business value, not testing events or tasks. Nevertheless, the team had been trained to create a backlog of user stories, allocate them to sprints, break them into tasks, and then manage the tasks visually in the scrum meeting. Jackson figured that that wasn't a bad way to manage the room, so he accepted this lack of scrum purity. The team had evolved this practice, putting all the major upcoming events into the backlog and using the sprint board and daily meeting to keep the broad team aligned. At the same time the software development was being managed in the subteams, some in more pure variants of scrum as Jackson appreciated, others less pure but no less effective.

Herb brought the team up to date on all of the tasks related to the integrating event. Here in week two of this sprint some tasks were in the Done column, several In Progress, and one remaining in To Do: Conduct the Event on Thursday. Herb reviewed the In Progress tasks, giving updates on the final touches and preparations, and shared his view that the event would be ready to go and successful. Herb saw no material obstacles in the way.

Pam in turn called on other participants with tasks one by one, moving quickly through the updates and either taking for herself or assigning follow-ups for impediments raised. Everything seemed to be moving along swimmingly, and she wrapped up her call for any other topics for the team that morning.

The team's newest member, Guptan Rajiputah, raised his hand. A bit unusual, hand-raising wasn't the norm, but Guptan was just learning the ropes. He had joined the team the previous week in charge of production support, reporting to Gary. The role had been called out as needed during the team configuration meeting and Gary had lucked out by finding him becoming available. Pam called on Guptan, who quietly began to ask his question.

"May I add a story and some tasks please, Pam?" Guptan asked.

"We usually don't add in this meeting, Guptan, we do that on Friday before each new sprint starts. But as long as we have a few moments, go ahead – what do you want to add?"

"I heard about the architecture simulation this team did and would like to request a similar meeting soon. I'm starting work on the production support plan and would like to build a base by identifying everything that could go wrong and how we detect and repair it."

Pam turned to Gary, since Guptan reported to him and Pam knew that Gary could figure out what to do with this request.

MEETING TIP

When a topic is raised in a meeting that doesn't quite fit, take it offline. Get it assigned to someone trusted to deal effectively with it.

"That sounds like a great idea, Guptan. Let's chat after this meeting, OK?" Gary responded.

Guptan nodded, Pam acknowledged, and the scrum meeting ended as the teams headed to their staked-out areas to get the job done.

14

Meeting or Systems Analysis?

BACKGROUND

A big, extraordinarily well-prepared meeting isn't necessarily the answer to every question. Big meetings are expensive, in time and often travel dollars. They can be tremendously valuable, as we've seen in the three big meetings so far – architecture simulation, project planning, and team configuration. But a big meeting is a hammer not suited to every nail. It is up to facilitative leaders to fit the right tool to the problem.

Should Gary accede to Guptan's request? Should Pacifica do another simulation soon, focused on laying the groundwork for production support?

Think about the need for a meeting in terms of the tenets of facilitative leadership:

- Rigor. Is a meeting the best way to get a rigorous decision? To make good decisions we need to build an information base, get options identified, and evaluate options based on facts and values.
- Alignment. Very often meetings are the best way to get alignment – i.e. ensuring that everyone with something to add has a chance to do so with appropriate facts at hand – and then enable understanding and effective execution.
- Efficiency. One way to think about the idea of efficiency is respecting people's time. It's a powerful way of thinking about Guptan's request: would holding a meeting of this type, at this time, be respectful of the time of everyone coming to it? Is there a better way?

- Frameworks. We've seen how powerful frameworks can accelerate RAE. When thinking about a need (like Guptan's), what is the appropriate framework? Is it the architecture simulation, or something else?

Let's see how Gary and Guptan talk through Guptan's request.

PACIFICA DECISION ON MEETING OR NOT

As the daily scrum broke up Gary walked over to Guptan and guided him to a huddle room off the studio to address his request. Gary was just getting to know Guptan; his manager, Heather, had assigned Guptan from another group in the Bank. Guptan came highly recommended for his technical and procedural skills in support engineering, a devotee of industry-standard change and problem management processes. Heather had alerted Gary that Guptan was somewhat socially awkward, introverted but curious, seemingly arrogant but in fact insecure in his communication skills due to a strong accent.

He started by asking Guptan to elaborate on his proposal.

"One of the foundations for building out our production support plan is doing the FMEA. But I don't know these systems or people so was excited to learn about the simulation approach. We could walk through each of the system flows and fill out the FMEA as we go, get it done in a few days."

Gary saw how enthusiastic Guptan was about this plan and was glad to see it. He had to admit, however, he had no idea what the "FMEA" was.

"Sorry Gary, on my old team everyone knew the FMEA." Unfortunately, Guptan said this in a way that sounded like "Sorry Gary you are stupid," but fortunately Gary had been warned. (Note – if you like Gary aren't sure what an FMEA is, check out the section in Appendix 2 on it, page 265. It's a powerful framework.) "It means Failure Mode and Effects Analysis. It is a standard engineering artifact. It identifies every component in the system and how it can fail, and then we add how we detect and repair the failures."

"I am thinking if we do the simulation again, we can stop at every point and identify how each step could fail and fill in the FMEA form. I could get the whole thing done well, quickly."

Gary really wasn't stupid, so he understood what Guptan had in mind. He was skeptical that it was the right approach, however. A few clarifying questions were in order but Gary had learned from Mary that starting off a critique by noting the positives first created a more receptive listener.

"Guptan, the idea sounds fantastic. A really organized way to lay a groundwork for production stability and support. I would like very much to learn more about the technique and help you get it done. I do have a few questions on the best approach to get it done. About how many people do you think we'd need to have at this meeting?"

"Perhaps 20 or 25 from what I can tell."

"Does everyone need to understand the failure modes of every component?"

"No, I don't think so. We just need a comprehensive central inventory of each element and failure mode so we can assess gaps and establish a support plan."

"Do you anticipate much interaction between the owners of the various components as they describe failure modes?"

"Perhaps some, such as when one system deals with a temporary outage by retry."

"Guptan, I appreciate how this meeting could be valuable to you – it would have everyone lined up to tell you in turn what you need to know. However, I don't see that it would be beneficial to the attendees – there is little interaction needed, they don't need to know the answers of the others, and they will spend a lot of time waiting or hearing things they don't really care about.

"Guptan," Gary concluded, "this is a problem better assigned to handling in one-on-one meetings with each component area and fleshing out the FMEA that way. Big meetings to extract data to fill in a form are a waste of time and disrespectful of the time of their participants. Perhaps once you get

LEADERSHIP TIP

Guptan's idea fails the RAE test. The meeting would have little impact on rigor, isn't needed for alignment, and certainly fails the efficiency test. We need RAE around production support, but a big meeting now isn't the right way.

a good draft of the FMEA it would be worth reviewing implications with part of the broader team."

Guptan looked disappointed but accepted Gary's recommendation. Guptan promised to meet with Herb to get a few milestones related to the FMEA onto the project plan. He also silently plotted to get the FMEA onto the studio wall as quickly as he had meaningful information on a form. This team, he thought, had a lot to learn about reliability engineering.

15

Demos

BACKGROUND

"Working Software over Comprehensive Documentation" is the second Agile Manifesto value. The primacy of delivering software is perhaps the strongest thread through the Manifesto. This value is followed by several principles in a similar vein: early and continuous delivery, deliver frequently, working software as the primary measure of progress, reflection (made more valuable by actually having something on which to reflect).

The demonstration (demo) is a pillar of the scrum methodology. It implements these agile values and principles. In scrum, the demo is the culmination of each sprint – the time set aside to show to the sponsor the software that has been completed in the sprint. This is a great idea and has proven its value now over a decade or more.

I am a big fan of demos. In fact one of my rules of thumb on the likelihood of project success is the content of status meetings. If the meetings are all about process:

> We'd like to report that we have completed and reviewed sixteen user stories but had planned eighteen. The risk review of the vendor was to have been completed last week but now looks like three more weeks out. Therefore we are yellow.

As opposed to software:

> Today we would like to show you the login and customer-needs-to-product-fit software. We are taking a slightly different tack on it than you may have expected because we found that we couldn't find a way to make it perform fast enough doing it all at once, so we have

a three-step flow. We think it's still an excellent experience but want
to get your views on it.

The project is probably in trouble.

There is a lot of literature and training on how to conduct a demo
and its role in scrum, which I will not repeat here. Instead let's consider
some leadership aspects of the demo within the project flow.

- Be sure you follow the RAE meeting guidelines. Have a clear
  objective, meeting path, and outcome defined. Too often demos
  are just a show and tell (still valuable but not as RAE as it
  could be).
- How polished a demo? In short sprint periods, say two weeks, it
  can take quite significant effort to prepare for a demo. This is
  a common debate – the value of the demo versus the time it takes
  to prepare for it. Remember that one of the goals of the demo is to
  focus the team on delivery – getting the code ready to show
  shouldn't be wasted time. Let RAE be your guide. (As an aside,
  the overhead to prepare and execute the valuable sprint ceremonies such as demo, sprint plan, and retrospective is one reason
  I tend to prefer somewhat longer sprints than the two-week period
  now in vogue.)
- Is a demo sufficient to accomplish your goals? Often one of the
  goals of doing regular demos is to validate that the product meets
  the expectations of the future user/owner community. To para-
  phrase a popular comedian: Have you ever spent a year on
  a project, delivered quality code meeting all stated requirements
  to UAT, been totally rejected by the owning business group and
  spent another year making dozens of changes and added features?
  I have. A demo can deliver on this goal, but it may not if it doesn't
  include all the potential objectors, is not in enough depth, or
  doesn't explicitly ask for this kind of approval and hold the
  approvers accountable in some way. Good leadership can avoid
  this risk.
- Documenting outcomes. Some agile projects take pride in not
  doing meeting minutes or status reports other than standard tool
  outputs like burn down charts and impediment lists. This gets to
  alignment and rigor and efficiency as well – does doing a demo
  report and distributing to attendees and those who cannot attend

help get sharper outcomes, better input and alignment, and is it worth the work? RAE.

You should now get the message – the purpose of the demo is to focus the team on delivery consistent with agile values and principles. As a leader the demo is a valuable ceremony you must guide to enhance the rigor, alignment, and efficiency of the program.

THE PACIFICA DEMO

The Pacifica team had set their regular demo to synchronize with the two-week studio sprints rather than their integrating events. The two cadences were not completely aligned so it was communicated and accepted that sometimes they would be able to show new working software and other times they would not. The integrating events, it was understood, was where in-depth review and understanding and critique of the evolving solution would happen, not the demos. The demos had become a regular large team meeting open to the broader set of partners, managers, and the future user community. Jackson as usual had some trouble with this evolution, as it was not orthodox scrum methodology, but he didn't have any better ideas and it seemed to be working for the project.

At Mary's insistence, Pam had agreed to have someone else be master of ceremonies for the demos. Mary believed that spreading out the formal meeting leadership tasks got the team more involved (alignment). It also helped with rigor and efficiency by allowing the formal meeting leaders to focus more intensively on a smaller number of facilitation tasks. Surprisingly, Megan Williams eagerly asked for the opportunity and was off to a good start.

Megan Williams had never been on a full-time product development team. She was an English major who had bounced around several freelance and employee roles writing manuals, advertising copy, and now

LEADERSHIP TIP

The Pacifica team has created integrating events to give much greater routine focus to ensuring completion of each major software milestone, taking the place of demo prep in many scrum projects. This also led to a change in the nature of the demos. Leadership adjusts for RAE.

had this more stable job in Pacifica corporate communications. She was smart and ambitious if somewhat unfocused heretofore from a career perspective. Megan had taken immediately to the agile training and become a team contributor far beyond her formal role. During the team structure meeting she had been designated by the change management workstream to be their representative on the central leadership team.

Mary had been impressed with Megan's eagerness and talent and was excited to see her growth on the team. Facilitating the regular demo meetings was an excellent chance for Megan to practice her

LEADERSHIP TIP

Previously non-technical team members can get a spark from being on an agile team. Fan the flame.

formal leadership. Mary strongly wanted her and the demos to succeed, so she had made time to mentor Megan in meeting planning and conduct, and each demo agenda was reviewed with the full leadership team at the start of demo week.

Megan kicked off the sprint demo by welcoming the full team and broader demo review group. For demos the studio population swelled as fuller teams, sales and marketing, operations, and senior executives gathered. Today both Sai and Heather had been able to join, which was somewhat rare. Sai was a regular but Heather rarely could fit into her schedule. The proceedings were also made available, as best the technology would allow, to remote visitors through collaboration software. After the welcome, Megan moved on to lay out the objectives, path, and outcome of the meeting, referring to visuals on the wall and a printed handout.

"We have three primary objectives for this 1-hour meeting today," Megan began. "We will first do a brief status report on the accomplishments and misses of the just completed two-week sprint, and what that implies for our November first-use target date. Next, we will demonstrate the customer upload of the hedge transaction set from one of the popular accounting systems, which was one of the focuses of this week's integrating event. Finally, we will do a brief design review of the new CurrX capabilities.

"We are aiming at three outcomes for today's demo," Megan continued, and turned the audience's attention to a list written on the whiteboard wall. "First, understanding of status and any follow-up tasks that come from that. Second, share the transaction set upload process we've just completed with the broader group here today and identify any follow-up items from that. For example, if anyone wants further demonstrations or has other concerns, now is the time to ask. As a team we are moving beyond this step in the product and each of you has an accountability to speak now or forever hold your peace. Well, not forever, but certainly until after the November release. We don't want to get tripped up later on insufficiencies in the upload process if we can find them now."

Mary thought this a critical point and wanted to reinforce that thought, so she did.

"I'd like to be sure everyone in the room understands and agrees with what Megan just said. The agile process we are following depends on building and testing the MyHedge product incrementally from start to finish. We are putting the product together and validating function in our integrating events, fine tuning and debugging as we go, and then building the next steps on top of what we expect to be a solid foundation. Now is the time to weigh in.

LEADERSHIP TIP

As a meeting participant you can add RAE by reinforcing concepts and by interjecting a test, as Mary does here.

"Could I get a read on extent to which you all understand and agree with this? Fist if you disagree, five fingers if you are fully on board."

"Wait a moment, Mary, if you don't mind?" Sara Okada interrupted. "One of the agile principles is something like 'We welcome changing requirements even late in development.' Isn't your admonition in violation of that principle? The product management and sales teams really liked this principle. In the past we would have to sign in blood on requirements and design documents and then go through a very burdensome change control process."

"Great question, Sara. That is one of my favorite principles but also one of the least understood. Its first meaning is that we need to establish a project process that brings the requirements and design much closer in time to the coding, and delays decisions until needed and to when the most possible information is available. With respect to the data upload, product, and sales, others have been giving input through the development process, even making changes as recently as this last week as a result of the integrating event.

"Its second meaning, I believe," Mary expounded, "is to recognize that big upfront requirements and designs are by their nature wrong to some degree for two reasons: the pure volume of decisions represented in a large set of requirements, and the fact that both the understanding of and the actual facts in markets change.

"Assume there are a thousand decisions reflected in a specification. Just natural human fallibility would imply some of these are wrong or incomplete. What percent? Five? That would be 50 change requests as they are understood in subsequent development and testing steps, or 50 problems released to customers. So, to have a great product, we have to welcome change requests, not discourage them.

"Similarly, if the market changes, for example a new accounting system has gained in popularity and we should support it instead of, or in addition to, the ones we have already specified, we certainly welcome that information and would like our project process to be able to accommodate that change. Our goal is to satisfy customers and make money for the Bank.

"Sara, let's consider this last example. Assume a new vendor was gaining share and we wanted to substitute its file format for one of the vendors we now support. In a non-agile project the vendor list would have been finalized months or more ago, and then would have proceeded to the design phase, and likely not coded and tested until a few months from now. Change would have been more difficult. In this project we would have welcomed change up until recently since we concentrated all the detailed analysis, design, coding, and testing into just a few weeks.

"However, there still comes a time after which change is expensive, and for file uploads that time is now. We all have an obligation to give our input now. We will continue to welcome change, but it will inevitably come at a cost."

Sara was satisfied with and endorsed this explanation. Mary continued to finish her proposed fist to five, got a lot of fingers, and handed the meeting back to Megan.

"Let's see, where were we? Oh yeah, on the outcomes. The last outcome," Megan resumed, pointing to the list on the wall, "is an understanding of the CurrX technical design and implications for our product and this project. We will present the design options, and choices we are facing, and implications. Again, we are asking for input, agreement or not, or any next steps. CurrX is heading into the meat of its development work and this sets the basis, and as Mary just described, as much as we'd like to change the basic design later that comes with a stiff cost."

Satisfied that objectives and outcomes were understood and agreeable, Megan had Pam and Herb give an uneventful project status and then moved on to the demo. Lilian (dot com), Yong (integration), and Vivian (MyHedge) together showed a customer persona logging in, selecting the product, and uploading a file from his accounting system. Yong showed how the file went through data standardization and wrote the data to Vivian's cloud storage in MyHedge, and then Vivian logged in to MyHedge and showed the data in cloud store. Because this followed the integrating event by just a few days, the demo was near flawless and drew the applause of the full room.

The final element of the sprint demo was the consideration of the CurrX design. There had been debate on the leadership team about bringing this to the sprint demo. Some contended that it was too technical and wouldn't be relevant to the audience. David, the lead architect, was the loudest voice to include it. He contended that there were real business implications to the choices being made and believed that an important element of agile was to bring the technical and business choices into a closer partnership. It's time, he argued, to share these kinds of decisions and spread the ownership and risks. Mary had stayed out of the argument until near the end when she endorsed David's view. There are multiple ways, Mary agreed, for TradeX to implement Pacifica's needs, and there are business decisions and trade-offs we are making that should be visible. Mary's concern was whether the demo or governance was the proper forum for this important discussion, but she went along with the team decision to bring to demo.

By the time the demo got to Dexter and Ivan there were only 10 minutes left in the meeting. They had planned on 20 so started off behind. Dexter, as the Pacifica lead for CurrX, set the stage.

"We spent three days last week at TradeX's offices working through requirements and designs and would like to share some important decisions with you. We would like to explain and confirm some trade-offs upon which we have jointly agreed. I'll start with describing the challenge we face.

"For MyHedge, we have some new characteristics that aren't well suited to the existing CurrX product. CurrX is optimized for the existing FX hedging practices. The current pattern is to deal with a modest number of large transaction sets with near-continual evaluation, notification, and optional automatic adjustments of hedge buy and sell. Our new pattern is a large number of transaction sets with a smaller number of transactions, for which continual evaluation and automatic hedge adjustment is not needed. I'll stop here and take questions – it's important that we understand and agree to these criteria, they drive important architectural design decisions we are making."

Sai, who had been quiet up to now, had his attention perked by something David said. "David, the decision on automatic hedge adjustment. Could you elaborate on that? I'm not so sure we don't want or need that in the future for these customers."

David deferred to Sara (product owner), who had laid out these needs with the team.

Sara explained, "Our persona are small business owners, most of whom are not sophisticated at hedging foreign exchange risk. Our customer research showed that if we were to automatically spend their money changing the risk profile of their future currency flows they would likely not understand or appreciate it. They want to be in control of any decisions to spend their money.

"It's worth talking about the risk evaluation periods as well," Sara continued. "Large companies are dealing with large dollar amounts for which even modest exchange rate changes can have big financial impacts, and they have the staff to monitor and manage risks continually. Small companies have much smaller impacts of fluctuations – an exchange rate movement has to be quite large to justify a change in hedge strategy. If we continually notified small business owners of change they would be overwhelmed and would likely miss the signal from the noise."

Sai was satisfied with the explanation and said he appreciated how the actual engagement with customers as part of the product development had revealed these differences. He wondered, though, what would happen as small customers grew to become larger? Would they have to switch products? Sara explains that yes, Pacifica would be offering two automatic hedge management products, and customers could choose which to take. Switching between them would eventually be automated but that wasn't in the MVP first release.

With agreement on the needs of the new MyHedge product, David asked Ivan to explain what it meant to the design. There were only a few minutes left so he hurried.

"Because of the different characteristics we are going to create a separate data store and evaluation process for MyHedge transaction sets. We believe this will be a capability desired by other CurrX customers, so we are making this feature part of our base product. The business implications of these decisions for you are," Ivan rushed, as he put up a summary slide, "that we need to go through our standard base product engineering and release cycle, and that we and you will need to establish this new data store and processing flows and confirmations in our production systems."

David wrapped the meeting up. "The alternatives were to do as custom for us and productize later, or to hack the current system to make it work for a small volume

LEADERSHIP TIP

If you plan on sharing an important decision such as this CurrX approach with the team for rigor (get input) and alignment, don't be satisfied with a half-hearted attempt. Here the team decided to share but accomplished neither rigor nor alignment.

pilot. TradeX recommended the approach and the leadership team accepted it, subject to this review."

By this time the top of the hour had come and gone, and the room was starting to clear out as people rushed to make their next meeting. To most attendees David's description sounded like a technical question and the leadership team had agreed, so there was little debate or discussion. This would come to be a topic of some heat in a future retrospective.

16

Governance Meetings

BACKGROUND

In one agile implementation in which I was involved, early on we reviewed, modified, and then published a set of principles we were seeking to promulgate. The principles were like the Manifesto principles but had some twists. The governance concepts in the Manifesto, to the extent there are any, seek highly autonomous and trusted teams.

- "Build projects around motivated individuals, give them the environment and support they need, and trust them to get the job done."
- "The best architectures, requirements, and designs emerge from self-organizing teams."

The principle we considered was light governance. Quite simple, but what does it mean? As a broad spectrum of executives discussed its meaning, one thing was clear – many of the executives felt that projects were over-controlled, that teams spent too much time and energy reporting on status and gaining approval for decisions, and that as a result they moved too slowly and did not take enough risks.

But what did it mean to have light governance? Did it mean, as one executive suggested, that teams report status and request approvals only as they choose to do so? That if executives wanted to know what was happening they had to attend the demos or drop in on scrum meetings? That managers should simply trust the teams to get the job done?

As I argued in the "Connecting to the Broader Environment" section of Chapter 12, beginning on page 162, proper governance helps us be rigorous, have alignment, and be efficient. In organizations of any size, there are leaders in the hierarchy above the team members who are being held accountable for the project and have the organizational power to help or hurt progress towards success. Often in my experience these leaders are not just empty suits – they often have knowledge, judgment, and relationships of real value to accomplishing team goals. A governance group can be an efficient way of ensuring rigorous alignment.

Let's assume for this section that we've agreed to establish a governance group for a program and we want to make it work in an RAE way. Let's just touch on a few dimensions that might be of use, and then turn to the next Pacifica governance team meeting. The topic hurried over in the scrum meeting we just saw, TradeX's choice on how to implement Pacifica's new product requirements, has reared its ugly head.

Cater to the "project owner." There are natural roles on many governance groups. Typically, one organization owns the program, usually whichever group is paying for it and will get the highest benefit from its success. In the Pacifica case, this is Sai's business unit. The leader of this unit is likely to be first among equals on governance, and governance meetings should cater to this leader. Project leaders should ensure their governance meetings deal with issues of most concern to this leader (or leaders), stage the meeting flow to provide information and opinions in a consumable way for this person, and make decisions/next steps in a manner with which this person is comfortable. This is not a good thing – it's a practical thing. Think for example of a meeting with the President of the US in the Situation Room; certainly, all attendees are valuable, but here one perspective matters more than the others and it would be foolish to ignore that.

Assemble the right members. Catering to the project owner(s) does not mean that other perspectives do not matter. In fact, often a good governance group comprises a subset of management hierarchy of the project leadership team. For Pacifica, the technology group reports up to Heather and she is invested in its success so she should be a member of governance. Some of Heather and Sai's directs or peers may also be valuable members, e.g. the leader of dot com, or an information security expert, or Sai's sales or marketing or finance leaders. The right people on governance enhances RAE.

One common debate I've seen over the years is whether vendors should be on governance. The Agile Manifesto speaks to this in one of its four values: "Customer Collaboration over Contract Negotiation." I suggest erring on the side of inclusion of all critical partners.

For topics and content, less is more. If assembled well, the governance group is a highly valuable set of powerful helpers/inhibitors, for whom any individual project is an important but not dominant concern. The members each have different foci, different experience and value frames, and different information coming into each encounter. Typically, the duration of each encounter is sharply limited as well, maybe an hour, rarely more. To efficiently use this powerful tool, the program leadership must limit topics and conversations to the most important items only.

Furthermore, senior executives are much more likely to want to hear a story from trusted experts than look at detailed diagrams and tables of data. They want digested views well-explained (a few compelling diagrams) and well-supported (summary data backed up and testified to by other attendees). I recently was in a steering group meeting in which the presenter turned to a page densely packed with project status updates and remarked "there is a lot on this page so I won't read it all to you, any questions or comments?" Either don't put that page in the report or know why you did and what you want out of it.

All of the normal meeting preparation tips apply. Explain upfront why you've chosen the topics on the agenda and what you are asking of the group. Be clear when you are providing information versus asking for a decision. When asking for a decision be sure this is the right forum and the right time.

The standardization dilemma. Imagine you are a senior executive who is on several governance groups, perhaps a few at a time which regularly turnover. It is natural to think that standardizing the information agenda, format, and content would make it easier to consume and engage with the material being presented. I've seen examples of this idea taken to bizarre extremes – one format required pages that must be included even if completely irrelevant, with the words "Not Applicable" consuming the full page!

I can see some value in standardizing formats of more voluminous and less personal communications such as status reports (although I'm not a big fan of that either), but strongly advise to standardize governance meeting agendas and formats only at a conceptual level driven by RAE leadership concepts. Something like:

- What topics are we covering today and why are those the most important ones? Often this section will touch on overall project status, saying most things are OK but we need to talk about topics to follow.
- Topic one. Provide enough background so the diverse group is well-informed, pose the choices available (if you want guidance or help), and set next steps.
- Topic two: Same.
- Topic three: Same.
- Wrap up.

For each topic provide the information and choices in the most compelling way you can. If you are fortunate enough to have an A3 culture, that can be a great format for each topic. However, if you don't have an A3 culture, that format can be too much to absorb on one page. Use conceptual diagrams, simple decision flow charts, options lists with pros and cons, decision or 2 x 2 matrices, but keep the concepts very few to a page. You want to encourage alignment and efficiency, so you need to keep the dialogue focused.

Enough background. Let's see how the next Pacifica governance meeting goes.

PACIFICA GOVERNANCE GROUP MEETING

It hadn't been easy to get a time that worked for all of the governance participants. Sai's assistant had used email and phone calls to get a 1-hour slot every other Thursday at 9:00 AM. Sai and Heather had both asked the invitees to make a special effort to attend regularly.

The team members attending governance included the entire leadership team, about 15 people, plus the executives. There had been debate about whether to include the entire leadership or just the top tier, ultimately decided by Gary and Sara who wanted to include the whole team, which proved acceptable to Sai and Heather. Sai was clearly the loudest leadership voice since he had the profit/loss ownership and was the driving business leader behind the initiative. Another strong voice was Heather, the technology executive to whom several of the tech leads reported. A new voice was Max Rosenfeld, the EVP of Product Delivery for TradeX whose portfolio included the CurrX product. Mary had advocated strongly to ensure that CurrX was well-represented on governance – this project was important to them and their delivery was foundational to MyHedge. A few other directs and peers of Sai and Heather rounded out the group of about 25.

Prior to the first meeting of governance Mary had guided Sara, Gary, Pam, and Herb in establishing the forum. They worked together to write a one-page visual proposal for how they suggested governance work and then met individually first with Sai and then with Heather to refine it. Their proposal as drafted was nearly acceptable to each of them, so with a few modifications in membership governance was established. The leadership team had agreed that Sara would facilitate governance meetings.

Governance had had several meetings so far and was starting to find a bit of familiarity. Sai and Heather were clearly the top decision-makers in the group. The meetings often paused for attendees to confirm items with one or the other. By now participants understood that the forum's attendance was quite broad, a place to learn and kick around ideas and set next steps but not necessarily a place for frank discussions better done in smaller groups.

For this meeting Sara was in a room with Gary, Sai, Heather, and many of the Pacifica team members. Ivan Onomarenko from TradeX was also present but his boss Max was on the telephone and Webex.

Sara dialed in early and made sure that Max and others didn't have a technical problem with connection, which seemed to happen at the worst times. She waited a few minutes past the hour as people gathered in the room and online and then kicked the meeting off.

"OK, we are ready to start," she began. She got through introductions and took care to lay out the agenda for the meeting. "In today's meeting we'll do three things. First, I will give you a brief overall status on the project, which is generally going well but has one looming concern. Next, we'll do a demo of a major piece of the solution, which we think you'll like. Finally, we'll dig into a concern with the TradeX product development work for our CurrX system at the heart of the hedging logic. We are not asking for decisions today, and we need help only on getting the contract with TradeX done quickly so as not to further jeopardize progress.

"We are generally on track for November delivery but there are still several uncertainties. We have accomplished a lot, as we will show. Our next agenda item is a demo of the process of buying a hedge product and uploading from an accounting system. Our focus now is on the back-end transaction set storage, the generation of change events, and the automated and manual reactions to change events. This is the area of concern which we'll cover after the demonstration.

"Customer research on user needs has been valuable and we are continuing to engage with prospective users for feature definition. We are increasingly confident that MyHedge has a large customer base eager for its features."

Sara checked in with Sai and Heather directly on alignment with the agenda and any questions on the status before proceeding. All good, so she turned the virtual floor over to do the demonstration.

Judy Grossman (User Experience) did the demo with occasional elaboration from Sara, Mark Flannigan (dot com), and Vivian Lockerbee (MyHedge). Yong (Middleware) had to answer one question regarding the data transformation. The demo went well and it was clear to governance that the front end of the product development was on an excellent track.

Sara passed the torch to Gary for the discussion of the back-end development. Gary had prepared a slide laying out the foundational decision for TradeX to build a new module for MyHedge due to the differences in product volumes, lack of intense real-time evaluation, and desire to keep costs as low as possible. Instead of using a relational database for persistence and relying on persistent large memory arrays for essentially constant real-time evaluation, CurrX was going to use a non-SQL database and pull data to memory only upon a set of triggers for evaluation. The triggers for evaluation would be a specified change in exchange rates, weighted by the volume of transactions for each customer. The system would also execute a nightly batch evaluation for every set, regardless of change in rates. The benefits for scaling, cost control, and simplicity of the product for customers was compelling, he said.

This sparked a series of questions from finance, sales, and Sai, and the team did their best to provide more details on projected cost savings in technical infrastructure and perspectives from the customer testing they had done. As Gary sensed alignment around the decision, he continued on to the concern he wanted to share.

"While we all agreed to make this a standard part of the CurrX product and to use a new technical architecture, getting this done on time has now become the critical path for the project. TradeX has limited skills and expertise in this new technology, as do we, and we both need to set up new servers and other infrastructure in multiple environments. We have worked hard together to lay out a plan that gets us to November but it has little slack in it. Any material slip could push us into next year."

Sai had already spoken to Max on this but wanted the whole group to hear Max's perspective. He asked for Max's comments.

"Gary has summarized the situation well. We did not have this on our roadmap for this year and already had a full plate. As your customer and product research reached conclusions earlier this year and the opportunity was understood, it took us a while to absorb and agree that there was opportunity for us as well. By the time our product management and technical teams had dimensioned the MVP with your team it left us little time to get this done.

"On the basis of verbal agreements we have added resources to the effort and reprioritized the roadmap as much as we can. We do need to have the Statement of Work signed within a week or we will need to put a hold on development. We could use your help to push that through, Sai."

Sai asked his finance manager for status on the contract and was advised to take it offline after the meeting. Sai assured Max that he'd give the matter his full attention and would contact him directly by end of day one way or another.

Heather had some additional questions for the assembled team. Was there anything else we could do to reduce the risk? Are the other teams building to solid interface specs and testing against mocked-up data and services? Could TradeX or Pacifica add resources to help? Could we do a pilot in the existing CurrX system without the new technology?

As the hour came to a close, it was evident to everyone that the outlook for November was uncertain. Herb closed the segment on CurrX by proposing that the project was now officially "Yellow," and Heather and Sai agreed. Heather told Max she would follow-up after the meeting, that she would like to come visit TradeX and get to know them and their plans better, and Max said he would be happy to host her at her convenience.

LEADERSHIP TIP

When the going gets tough, double down on in-person relationships.

Sara closed the meeting on a somber note, but with the issues understood and Sai, Heather, and Max mobilized to help.

17

Teleconferences

BACKGROUND

The Pacifica governance meeting we've just seen was a combination of in-person and remote participation. Because so much of today's interactions are among geographically dispersed team members, bringing leadership to these virtual meetings has become an important quill in our leadership quiver. In this chapter I'll provide some tips, then we'll move to see an important teleconference in the Pacifica project.

A favorite Manifesto principle is "The most efficient and effective method of conveying information to and within a development team is face-to-face conversation." I could not agree more strongly. Despite improved tools such as Webex and Slack and video conferencing, Jira and email and Sharepoint and wikis, nothing has come close to providing the sheer communication bandwidth of in-person communication. Leaders should do everything they can to exploit this fact: co-locating teams, holding extraordinarily well-prepared in-person meetings, and allocating sufficient travel time and budgets. Like the agile values, however, the preference for in-person communication does not mean that other communication has no value or should not be done. Which brings us to group phone calls.

The challenges of group phone calls are well-understood by most of us who regularly need to attend them. Multi-tasking, awkward interruptions and talking over each other, inability to get visual cues on alignment or disagreement, lack of collaboration tools such as a whiteboard, technology challenges, irrelevant noise (the infamous toilet flush or dog bark); all contribute to the difficulty of group calls.

Here are some tips to make the best of a sub-optimal format for collaboration.

- Do a good agenda with meeting objectives, participants, path, and expected outcomes, and cover that first in the call. As in any meeting, be sure the attendees match the objectives and outcomes sought, and that preparation of people and material is extraordinary.
- Keep the agenda and the flow as simple as you can. With lower communication bandwidth, transmit less information.
- Try to take the technology flaws out of the equation by starting the conference session a few minutes early and ensuring that anyone using the tech, e.g. taking over a screen to present, knows how to do that.
- When folks join early, chat a little on innocuous topics like the weather or vacation or sports. This avoids dead silence and helps build human connections that otherwise are difficult over the phone.
- Start as soon as the start time has arrived AND the key decision-maker(s) are present.
- There must be a strong meeting conductor. This ensures RAE; without it, the path will be unclear, rigorous discussion will muddy, and inefficiency will reign.
- The conductor must conduct. That means calling on important participants directly for their views, asking people to hold off-topic remarks for later, and taking inappropriate interactions offline. Such inappropriate interactions could include newly introduced subjects for which the audience is not prepared; sensitive decisions and discussions best done more privately; heated dialogue that needs some cooling down.
- Ensuring alignment is much more difficult on the phone when you cannot see people's faces. The conductor or even some participants need to check verbally for alignment or even for assent to move on to the next topic on the agenda. One trick I've seen work well is to designate a leader for each room containing several people, especially if one of them is the room with the top project sponsor. The room-sitter is responsible to watch and help with interactions.
- Meeting material must be tailored for the format. Make the material line up directly with the agenda, don't include material that isn't absolutely essential for the dialogue (i.e. put details that

are unlikely to be reviewed in the call itself into an appendix), and use diagrams and conceptual slides more often than you might for in-person discussion.

We've seen some of these techniques used in the governance meeting preceding. Now let's see another as the Pacifica team deals with the project disruption arising from TradeX's architectural decisions.

PACIFICA TELECONFERENCE: USE OF AN A3

A month had passed since the governance meeting highlighting the risks in CurrX. Summer had blossomed in California and the heat outside was rising along with the heat on the team. The original plan had called for the CurrX base product to be code complete at this point and in TradeX testing, while enough of the other components were to be in the newly created end-to-end test environment so that a data test from customer to CurrX could be conducted.

The good news was that the original plan's milestones were nearly complete. Complete enough, indeed, that the two sets of formal testing could begin so long as they worked around some gaps being filled over the next several weeks. The bad news was that serious deficiencies in the CurrX system's performance and scalability had been found.

When Ivan (CurrX Product Manager) had raised the concern to Dexter (Pacifica CurrX tech lead) last week, Dexter had immediately alerted Gary. Gary and Mary had been conferring when Dexter interrupted with the bad news and the three of them discussed how to approach the concern. Mary suggested that Dexter and Ivan put together an A3 problem document and then get the technology team together to understand the problem and potential reactions or solutions. Neither had done A3s before so Mary explained it as a simple communication vehicle derived from Toyota's lean practices that helps in rigorous decision-making. (See Appendix 2, page 257 for A3 description.)

Over the next few days Mary helped Ivan and Dexter put the A3 together. Ivan started with a statement of the problem, which the three of them considered and adjusted and then Ivan reviewed with Max, while Dexter ran it by Gary. Mary helped them convert a lot of words into a few compelling diagrams and tables. They proceeded to document the current state of the technology and the implications for the project, followed by some options to address, and arguments for and against, each option. Again Mary counseled them to stop and socialize the information to be sure the A3 reflected not only their understanding but also that of their bosses. Each also conferred with other colleagues as they needed to.

With the time for a decision on the path approaching quickly and the A3 draft complete, Mary counseled that the next step should be for the technology teams to discuss the A3 and see if a consensus on the best option emerged. This would ensure that when the situation was elevated to governance there would be no debate or confusion on the facts of the technology itself, leaving focus on trading off risks and benefits to best advantage the Pacifica business.

LEADERSHIP TIP

Just because we are tearing down silos it doesn't mean that technology-focused team members can't or shouldn't have their own meetings or forums. If it advances rigor, alignment, and efficiency it's the right behavior. The same applies to each sub-team.

Wednesday afternoon arrived and with it the teleconference review of the CurrX performance concern. Gary decided to lead the call and use the A3 as the primary visual aid. Joining Gary in the room at Pacifica were Herb (Program Manager), Dexter (Pacifica CurrX tech lead); Janice (Pacifica CurrX Release Manager), Vivian (MyHedge Development Manager); Damien (Test Manager); and David (Architect). Gary had invited his boss Heather; she was in Boston so was calling in. From CurrX Ivan was in the room with Gary and the team, but Max Rosenfeld, the TradeX EVP, and a few other CurrX engineers were on the phone, some in a room at headquarters but others not.

A few minutes to the hour Gary opened the conference call line and the screen sharing application on his notebook. Heather was the first to join, unusual for her, and they compared notes on the weather and Heather talked about the daughter she had visited the prior evening. Once Max joined a few minutes later, Gary immediately called the meeting to order.

"We are here today to discuss the CurrX concerns for our November ship date. Dexter and Ivan have put together the problem analysis for us, which was in the email and is now on the screen share. Does everyone have that document?"

Moving on, Gary sought to confirm the outcome of the meeting. "Our goal today is to have a clear story and if possible a recommendation for governance next Monday. Max, does that make sense to you? Are we ready for that?"

"Yes we have enough information and have a direction we'd like to recommend," Max confirmed. "Ivan is prepared to walk us through the analysis document now."

MEETING TIP

The teleconference conductor should directly hand the floor to the appropriate leader at the right times.

Ivan directed the attention of the meeting participants to the A3 the team had prepared on the performance problem (Figure 17.1: CurrX Performance Issue A3 is an illustration of the A3 being discussed). "Let's start in the first box on the top left with the problem description," he began.

"We are now code complete with the capabilities that will support MyHedge and the good news is that it works well – for a small number of transaction sets. However, when we started performance testing we found that when the volume of transaction sets rises, we get a performance hit not only on the new MyHedge transaction sets but also on the existing large-volume and highly sensitive CurrX workload. You can see in the 'Current State' box the chart showing how both workloads slow down with only modest growth in the MyHedge volumes. Here we are showing CPU consumption as MyHedge transaction sets grow."

As meeting conductor, Gary worked to keep the flow moving, periodically checking in with Heather. Heather hadn't been engaged as deeply in the details of the CurrX problem as most of the others on the phone, so if the meeting was working for her it was likely working for all. Gary set up the next part of the dialogue by asking if we knew the cause of the problem.

Ivan continued his exposition, asking the meeting participants to refer to the diagram on the bottom left of the A3. "Our design approach was to create a separate data store for the new capability and a triggering mechanism to load the data to memory when we needed to evaluate it, but to use the existing CurrX core capabilities to do the

Pacifica Bank MyHedge Program: CurrX Performance Issue

Problem Description

- As number of hedge sets grows CPU consumption on base CurrX rises quickly
- This is an architectural issue we cannot fix quickly

CPU Consumption

Current State

- As number of hedge sets grows CPU consumption on base CurrX rises quickly
- This is an architectural issue we cannot fix quickly 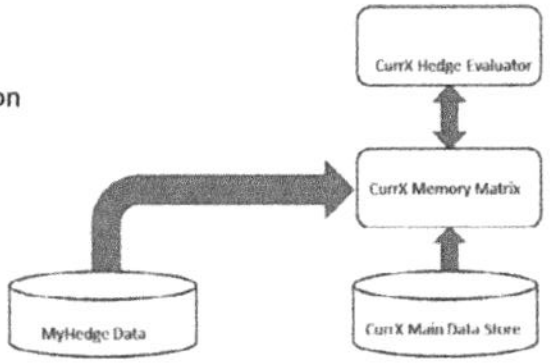
- The fix is to completely separate MyHedge at runtime
- TradeX estimates code complete on this not until end of Q1

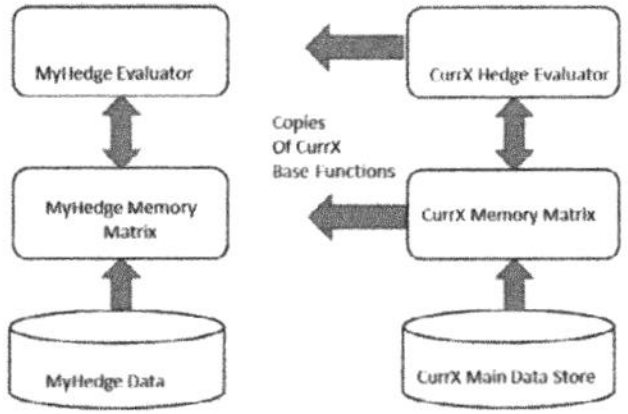

CurrX Hedge Evaluator
CurrX Memory Matrix
CurrX Main Data Store
MyHedge Data

CurrX Hedge Evaluator
CurrX Memory Matrix
CurrX Main Data Store
Copies Of CurrX Base Functions
MyHedge Evaluator
MyHedge Memory Matrix
MyHedge Data

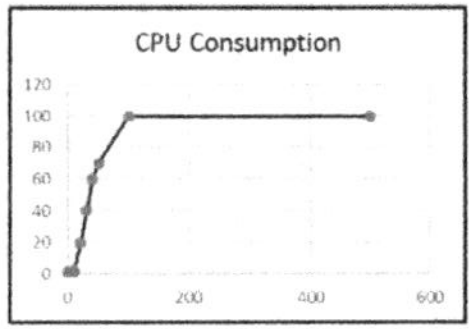

Options

1. Delay pilot until re-engineering complete
2. Extended low-volume pilot while re-engineering
3. Add hardware capacity to enable more volume while re-engineering

Analysis of Options

| | | Pilot | Full Volume | |
Option	Pilot Begins	Customers	Enabled	Expense Add
1	Next April	Unlimited	Next April	$400k
2	November	~10	Next June	$800k
3	November	~50	Next June	$1.2m

- Expense estimates are plus or minus $100k
- Customer estimates are approximate for planning purposes
- Delay in full volume in options 2 and 3 is due to need to support temporary pilot

Decision / Next Steps

FIGURE 17.1
CurrX performance Issue A3.

evaluation and manage the output. This would be the simplest approach, quickest to market, and easiest to maintain. You can see the basic design in the top diagram.

"We were surprised to see that the process of loading the MyHedge data to the CurrX memory matrix consumed a lot of CPU, and as the number of transaction sets grew the simultaneous triggering of evaluation wound up being a major bottleneck. Worse of all is the impact on the core CurrX system itself – MyHedge's demands for processing power essentially bring the overall system to a halt."

Dexter now added, "This would be deadly to our Pacifica operations. It could interfere with all our existing business. We could possibly throw more CPU at the problem but it's a lot and we'd need to upgrade our servers in all our environments, which would be costly and take some time. We really need to deal with software architecture."

Ivan confirmed Dexter's conclusion. "That's right, Dexter. This is an architectural issue that can only be fixed with some major surgery. The diagram on the bottom left illustrates what we need to do: we need to completely separate the MyHedge workload from the core CurrX work-flow by replicating the memory matrix and the evaluation routines on separate compute infrastructures."

Gary did another check to be sure everyone understood the problem. Heather had one more question. "Ivan, why didn't we foresee this issue when we did the design?"

Max decided to take this one. "Heather, I'll take that one. This was a new architecture for us and we didn't understand the added load on the CurrX processor. It turns out that bringing data into memory on the CurrX database server is much more demanding coming from an external source than from its own data stores by an order of magnitude. We have reviewed the problem with experts, and they agree the problem is entirely due to our design. We made a mistake, I'm just glad we caught it now and not after we went live."

Gary summarized before moving on the options. "So, we made a mistake in the design, and the result is that any meaningful volume of MyHedge transaction sets poses a danger to the operation of CurrX for current users. Let's consider the options, shall we? November is not very far away and we're supposed to be code complete now."

MEETING TIP

Gary as conductor of the teleconference is ensuring that each step is summarized and agreed before moving along.

Ivan took back over the reins and directed attention to a table at the top right of the A3 page. "We've identified three options, combining how to fix the problem with impact to the project. From the CurrX product perspective, we have concluded that we can't risk deterioration of current system performance by adding a new module to support MyHedge-like workloads. We are therefore going to completely isolate the new business from the old, while duplicating and synchronizing key features such as foreign exchange rate feed, the hedge calculation engine, and notifications. Our best estimate of when this may be complete is unfortunately next January or February – too late to meet the Pacifica launch goal."

Gary clarified, "Ivan, is that a firm decision? No wiggle room?"

"Absolutely, yes, Gary," Ivan confirmed. "We cannot risk the existing business and won't release a product that could harm our customers.

"That is option one: we re-engineer the MyHedge capability and the project is delayed until we can get that into production at Pacifica, likely April next year, roughly a five-month delay to first use and full production capacity. There are some additional hardware costs for Pacifica to isolate the features as well. The box in the middle right of the A3 summarizes the options. You can see estimates of pilot start, pilot volume supported, full volume capability date, and incremental cost increases in the table there.

"Option two is that while we are re-engineering the MyHedge capability we go ahead and get the current flawed architecture into production at Pacifica but limit the volume to well below the risk threshold, and we monitor it very closely. We can still meet the November launch date but a low-volume pilot period would have to be extended until the new product architecture is ready. Because we would need to devote resources to finishing and supporting the temporary pilot system, in this option full volume capability is probably not until June next year. There are also additional expenses in several areas to support the interim solution while building and testing the permanent fix.

"Option three is similar to option two but Pacifica adds a lot of hardware capacity to your CurrX environment, which we believe would enable you to support a larger pilot. The costs shown are what we think might be reasonable, and the larger capacity as estimated is shown as well. In no case would we jeopardize current customers, so we would monitor closely and be prepared to end ramp-up or even end the pilot if necessary."

With the options identified, the meeting continued considering the arguments for and against each, until Heather, Gary, Max, Dexter, and Ivan were in general agreement – options one and two were both viable, and it was a business trade-off decision to choose among them. That was for Sai, and it was needed soon.

Signposts	• In this section we have cruised through the set of meeting topics in which a project team would typically engage: demos, scrums, governance, non-meetings (when to avoid a meeting), teleconferences. For each we provided background and tips and showed the Pacifica team in action to demonstrate. • The MyHedge team has progressed from its early formation, made overall excellent progress along its plan, but has run into a serious barrier in execution. The CurrX product development team made an architectural error that cannot be easily corrected.
Leadership Guides	• Relentless focus on rigor, alignment, and efficiency shines through all of our work, ideally every day. This leadership approach improves each engagement and evolves the standard scrum method to fit our team and project. • Simple and attractive Agile Manifesto principles such as "self-managed team" give way to more nuanced arrangements, optimally connecting the team to the broader organization.
Coming Up Next	• We have now seen the broad array of leadership techniques and will explore the final major item on our plate: team reflection and commitment to aligned improvement. The next chapter explores RAE retrospectives. • Following the background chapter on retrospectives we'll see a milestone MyHedge retrospective and learn how the project turns out.

Section 4

Project Retrospectives

- We are now near the end of both our examination of leadership for agility and of the MyHedge project at Pacifica Bank.
- This section will examine project retrospectives. Reflection and adjustment is a part of all modern process management methods including lean, agile, and even military activities (after action reviews).
- The next chapter is our background examination of retrospectives. It includes a fairly lengthy example based on a past project of mine, which I call the "Saga of the Checkboxes."
- The second and last chapter in this section is the Pacifica project retrospective. In this chapter you'll learn how the project has proceeded through its November target go-live date and into the new year and observe an interesting retrospective technique in (fictional) action.

Following this "Retrospectives" section, there is a short summary chapter and then I leave you with a reference Appendix of common tools.

18

Background

Retrospectives

RETROSPECTIVES ARE AN IMPORTANT ELEMENT OF ALL AGILITY

Project retrospectives are a crucial part of leadership for agility. As we've discussed leadership techniques, one thing we see repeatedly is checking with our team and partners on alignment around path and process. We need to get input, evaluate facts and opinions, and ensure our team is working with common, same objectives and assumptions as they make their myriad daily decisions. The retrospective meeting is simply a formalized, regular meeting to ensure this happens.

Reflection has been a staple part of lean and six sigma for decades. The idea was embraced from the very start of formalized agile, as stated in one of the principles in the Agile Manifesto: "At regular intervals, the team reflects on how to become more effective, then tunes and adjusts its behavior accordingly."

The best teams have a common understanding of both the past and commitments to the future. Team members should have similar answers to what has gone well and what has not; similar answers to why things have gone well or not; and similar answers to what the team is emphasizing for the future in order to improve. Retrospectives aim to accomplish this alignment, rigorously and efficiently.

The set of techniques and leadership principles in which we have been immersed so far apply well to retrospectives. These include how to plan and conduct a meeting, how to efficiently achieve rigor and alignment. I won't go into the many techniques for doing retrospectives here – that has been done well by others, most particularly in *Agile*

Retrospectives: Making Good Teams Great by Esther Derby and Diana Larsen (Pragmatic Bookshelf, August 5, 2006). Rather, here I'll comment on just a few of the key leadership issues in retrospectives.

LEADERS ON THE TEAM: GETTING THE RETROSPECTIVE PROCESS RIGHT

At the moment, many agile scrum teams are choosing to do two-week development sprints. Within this period the team typically needs to do the detailed sprint plan and complete a set of user stories. The definition of complete varies somewhat but typically includes completing remaining detailed requirements and design, coding, testing, and debugging. The sprints would then end with a demonstration of the accomplishments, a retrospective, and transition to planning the next sprint.

This leaves little time for reflection of any depth or quality. Accordingly teams will often do more extensive reflections on a longer cycle, say every quarter, or after major release.

As a team member, your leadership obligation is to help the team efficiently conduct a rigorous examination of results and process. This means ensuring a proper framework is established onto which input can be hung and examined, and conclusions can be explored and ultimately adopted by the team. This is where *Agile Retrospectives'* suggested team exercises can be of great help.

In choosing the right framework, a couple of tips to keep in mind:

- Time. Is it possible to do a good reflection in the typical 1-hour-or-less event available in each sprint? Or do you need to do a super reflection event (my term) each month or even each quarter to go in much more depth?
- All together or enable individual feedback? The alignment requirement demands that we find mechanisms to enable input. Some team members may find it easier to contribute to a written survey or a 1–1 interview.
- Who to include? The tip here is to be true to the goal: a working team aligned on the past and sharing improvement commitments for the future. Most often this would exclude senior executives not deeply

engaged in day-to-day work but would include everyone regularly and importantly interacting with committed team members.

- What roles in the reflection process? Often the scrum master and/or the agile coach will take the lead role in driving retrospectives. It can be difficult to both facilitate and participate, so sometimes an outside facilitator can be useful particularly for large-scale milestone reflections.
- All together or some silo reflection? Particularly when the scrum master isn't strongly technical and there have been technical issues, engineers may want to either take a stronger role or even potentially do a technical retrospective without some of the broader team. This could apply to other areas as well. Don't be afraid to do work in silos so long as you leave the bridges open.
- How to communicate and implement results. End reflection with plans on how to execute the improvements that emerge. Don't over-focus on this; sometimes just having the team agree on change is enough and execution will evolve in the natural course of events.
- The framework/meeting path. Leverage good ideas from others but don't be afraid to put together a novel approach. As RAE skills grow there is plenty of room for innovation.

There can be dramatic power in a well-done retrospective. One of the most powerful I've seen was facilitated by a team member who volunteered for the mission out of frustration with the troubled project and a passion for success. Her first step was to speak individually with several of the team members, after which she put together a novel approach focused on emotionally compelling or memorable events and turning points. She limited attendance to the core team, no senior executives.

The approach was like the one Pacifica uses in the next chapter. It began with constructing a timeline of important decisions. Someone wrote a sticky that said, "We decided to pursue the ideal customer experience" and stuck it up near the inception of the project. Amazingly enough, this sticky got the most negative dot votes by far! The dialogue was nearly unanimous: that decision had resulted in a full year going by with nothing delivered and nothing practical underway. The team resolved to focus now on useful things they could deliver with less risk more quickly, while staying on architectural strategy as much as they could. In this case the "perfect" was the enemy of the "good."

This is the power of a compelling retro – the team arrived at a common understanding of the past, and from that understanding drew out a better future. In this case, no cookbook was followed, a team member not normally in a formal position of leadership took the initiative, and rigorous alignment was efficiently achieved. This is leadership for agility.

ORGANIZATIONAL LEADERS: INSIST ON ROOT CAUSE THINKING

The primary accountability of organizational leaders with respect to retrospection is to encourage RAE, accept and act on results if needed, and to insist on quality. The first two accountabilities are common requirements for organizational leaders across the full project spectrum. Insisting on quality has a special meaning with respect to retrospection because teams can, in the interest of maintaining harmony or through lack of rigorous thinking, quite easily settle on superficial improvement items rather than getting to the root of the issues.

To illustrate this trap and how management outside the team can help, let's consider the "Saga of the Checkboxes." This is a long story based on an actual delivery project that is rich in opportunity for learning.

The team was working to deliver electronic signing of a document package. Team members were new to agile thinking and quite comfortable in the waterfall world. Accordingly, the team started by having one of the business analysts write up requirements – what documents needed signing, by whom, at what stage of the business process.

One of the technology managers in a supervisory role had done electronic document signing and agile before and feared the consequences of incomplete requirements done in a technological vacuum driving the development process. She knew that the devil is in the details and wanted to get the team to get to brass tacks as quickly as possible, not waiting until the formal acceptance test at the end of the project to find gaps. She convinced the team to quickly stand up the system and get a package signed, to prove the system was workable and flesh out any issues.

The product manager and the scrum master worked together to do an early demo. The document team created one signing-ready package, UI/signing team configured the facility, and the stage was set. The product manager assembled, via Webex, a group of subject matter experts in operations, customer support, business analysis, legal, and compliance, and a demonstration was held in January. The happy result was agreement all around that the technology worked well and the team could proceed towards delivery by simply building out the rest of the document package variations and debugging at the end.

The committed and expected production date was May, and confidence was fairly high. May came and went, and full production didn't occur until almost a year later! The delay that ultimately became known as "The Saga of the Checkboxes" and the reflection that ultimately was done to understand and react to it is fertile ground for understanding the role of organizational leadership in retrospectives.

The first indication that the May date wouldn't be met came in March. A new package variation had been generated and handed off to the signing configuration analyst, who reported a problem. The package contained a document that had checkboxes from which the signer was required to select. The signing system was not designed to do anything other than guide the signer through the documents, stopping at each required signature line, and enabling the user to electronically sign the presented line. It had no capability of allowing the user to enter data. The implicit architecture was that all user interaction to make choices was to be in systems farther up the business process.

This set off a flurry of deeper analysis of the documents and additional problem instances were found. There were several text boxes, some required and some optional. There was a radio button display with contingent logic: if the user selected option b, a text box with details about the choice was required.

The team addressed these gaps with urgency but not much transparency. They decided that rather than dealing with the upstream systems to ensure that by the time of signing all user intent had been gathered, they would modify the signing system to handle these edge cases. This turned out to be more difficult than they initially thought, and more delays ensued. When the code dealing with the simplest case of optional checkboxes was complete, a demonstration was held, going in much more depth and with a more expert audience than the January demonstration. This time several more showstopper issues were found, creating even more delays. One of the most serious was that dealing with multiple signers in a myriad of potential configurations had been nearly ignored. The team stumbled forward, addressing each issue within the existing architecture and broad design, ultimately bringing an adequate solution for a subset of use cases to market about a year late. Another year of improvements would ultimately be needed to get the whole solution in place.

In June of year one of the Saga of the Checkboxes, the team did a focused retrospective on the failure to meet the May commitment. A few of the team members were gathered in a conference room, others were on the phone in far flung cities. The meeting didn't have a formal facilitator nor a planned structure; it was just a meeting to talk through what had happened and how to improve. The communicated outcome was an agreement that the problem was that the team was trying to do too many things at once. We either had to add staff to the team or be less ambitious.

When this conclusion was presented to the manager of technical members of the team (let's call her Lois), it was accepted as true as far as it went. But it didn't go nearly far enough – i.e. Lois exercised her leadership accountability to evaluate the retrospective quality and push the team to get to deeper and less comfortable conclusions on the failure. She asked:

- Why did the initial January demonstration miss the checkboxes?
- In solving for the checkboxes, why did the team change the foundational architecture (signing equals agreement, not data entry)? That was, in retrospect, a serious error that led down a rathole of system hacks. Was that understood? Were the right people and review involved in that decision?
- The team stumbled from issue to issue resolving each incrementally. Did the team do enough upfront analysis and detailed planning?
- Some of the delays were sparked by the first working checkboxes demo. Some key observers pointed out serious gaps in the user experience. They said that the team was delivering a document signing experience, not an "I understand and agree to these terms" experience. Why did the team mis-conceive the user experience need?

Working through her people on the team and her relationships with partners, Lois encouraged further conversations. Focused discussions on the architecture and path forward now that a key principle had been breached were conducted. Resource level options against alternative roadmaps were created and presented to executives. Partner team capabilities and attitudes were examined and debated. Over the course of that summer and fall, the implications of the Saga of the Checkboxes failure were more fully understood, and changes introduced that dramatically accelerated delivery and improved team satisfaction. These included:

- Yes, the team was trying to do too much with too little. That part of the team's initial retrospective was indeed correct. After consideration of alternative roadmaps and investment levels, the team was grown in several areas explicitly aligned to a reasonable delivery roadmap.
- Understanding the failure of the initial demonstration led to several changes. Part of the issue was ultimately understood to be lingering division of the business from technology left over from the pre-agile days. The demo had been assembled by the product manager and did not include the technical leaders who understood the architecture and details of system capability. The intended purpose of the demo was not rigorously aligned: technology wanted to take out technical risks while business wanted to minimally comply, on their own, with technology's demand for a demo.
- Examination of the technical workaround decision led to greater transparency of important design decisions and a commitment on the part of the technologists to push back instead of just acceding to immediate wishes. Everyone has to defend the long-term quality and integrity of the systems, it was agreed; we can hack together a work around, but only so long as we all understand and agree that is the right path.
- The stumble forward conclusion led to much more focus on detailed project plans and monitored commitments on tasks. The team added a junior project manager to help put cross-team plans together, monitor progress, and communicate issues to organizational leaders.
- The failure to understand multi-signer scenarios pointed out a gap in detailed systems analysis capability and failures in the requirements-to-design-to-code flow. This new-to-scrum team had done user stories with acceptance criteria in their agile management tool but never dug deeply enough into what was really needed. The multi-user confusion led to a very detailed multi-dimensional excel-based matrix of possible combinations of signers, types of signers, when the signer became known, package types, and other factors. The team had fundamentally misconstrued the user story

method for an excuse not to do detailed systems analysis – partially due to a lack of an explicit systems analyst role in the scrum model being used.

• The late realization that the entire user experience had been misunderstood to be about signing a document package instead of about understanding and agreeing to terms was perhaps the most impactful insight of all. This contributed to a movement to add customer co-creation to the development process, through which a user experience team member was added and regular customer/user testing was made a standard part of the development process.

This very extensive set of changes didn't all come just from Lois' insistence on a better retrospective. But Lois' refusal to accept the easy answers coming from the team was the spark and sustainer of the dialogues that led to dramatic improvements in market delivery.

We can learn from our failures and our successes, but it takes leadership to ask the hard questions in a way that gets to root causes and drives improvement. Lois' leadership is a good example of driving rigor and then alignment in retrospectives. This wraps up the Saga of the Checkboxes.

On to the Pacifica retrospective.

Signposts	The best teams have a common understanding of both the past and its commitments to the future. The retrospective is the formal way we seek to develop this.
Leadership Guides	• Teams will often do pro-forma restrospectives per scrum methodology, which may not add much value. Skilled facilitative leaders will help increase the value of reflection by supplying valuable frameworks. • Deeper-dive retrospectives at major milestones (major release instead of just each sprint) allow more time and perspective. • The organizational leader has an accountability to insist on deep root cause thinking. The "Saga of the Checkboxes" is a story that emphasizes the need to avoid simple surface explanations.
Coming Up Next	The Pacifica project completed its first release and is about to undertake its first full release retrospective, prior to celebrating the limited success.

19

Pacifica

Project Retrospective

January came to the studio, and with the new month and new year came the first birthday of the studio. Pam (Scrum Master) remembered the precise date and made sure the team was going to mark it. She nudged Sai who enthusiastically agreed to a celebration event. First, they would do the first major release-level retrospective, then cocktails and pupus on the lawn, where the tables and balloons were now being set up.

The past several months had been a whirlwind, followed by a few quiet weeks at year end. MyHedge was in pilot production now, with two pilot customers managing four total transaction sets. Two more customers were planning to begin pilot in the coming week following an incremental improvement release this weekend. The team had learned much and was excited for the June planned ability to expand the product they were now perfecting to a much broader customer base. There was a minor undercurrent sense of failure to deliver as much as they had been expected to/promised to/planned to/hoped to. This undercurrent was not mentioned often and hadn't dampened the overall enthusiasm for the studio and the positive results with the pilot customers.

Months ago, governance had been presented with some choices on how to respond to TradeX's inability to deliver the full solution on time (the situation is presented in the Teleconferences section beginning on page 225). The decision had been to go ahead with the functional but not scalable solution for November; to extend the pilot period until the scalable version of the software would be ready midyear; to perfect the product during that time; and then to go on a full-scale sales and marketing campaign confident in a market-fitting, well-working product.

The less-scalable version of MyHedge had been deemed good enough to begin a pilot roughly on time in early November and the initial user, a small importer in Bangalore who had been a long-term Pacifica customer, had been the first test in production. The set up and initiation of a hedge had gone smoothly, with no lack of close support (the sales person and Sara herself had been in the customer's office for the first login, captured in an iphone video, and several photos now adorning the studio wall). There had been a few hiccups in the triggering of hedge evaluations and notices back to the customer via PacificaBank.com, and the customer had some awkward interactions that led to some of the enhancements now going live. Another customer had joined in mid-December, more smoothly than the first, and was again deemed successful enough to add the next two soon.

As the team looked ahead, there remained a lot to do. They had to complete and test the new base version of CurrX, install the new infrastructure at Pacifica, install and test the new CurrX version and retrofit to the existing system without breaking current production. There was a long backlog of improvements and extensions stacked up. There were marketing and sales campaigns to plan and execute, a large-scale production support operation to get up and running, and personnel changes to absorb on the team. It was time to take a measure of what they had completed, learn from it, and make whatever improvements were needed as they headed into the next stage of the MyHedge program.

Pam and Jackson together wanted to lead this first major retrospective. Over the past year they had learned a lot from Mary on how to plan and conduct what Mary called extraordinarily well-planned meetings. Both realized that it was time to do an in-depth review, far more extensive and they hoped insightful than the brief, now pro-formal retrospectives done every two weeks in the studio. They had taken a stab at an agenda and then worked with Mary to refine it. It was now time to bring the agenda to life.

As the large wall clock tick-tocked 1:30 PM, Pam called the meeting to order. There had been much interest in the retrospective and turnout was high. About 40 people sat in their seats at the small tables distributed around the room, ready to give each other their perspectives on where the project had gone well and poorly, and on what to do better going forward.

"Welcome to our first full MyHedge release retrospective," Pam began. "As you know, we are doing a much deeper dive today than in our usual every two-week sprint retrospectives. Our objective today," she continued, as she flipped over the Welcome! sheet on her prepared flip chart to reveal a page titled Objective, "is to look back over the full last year and identify what we did well and where we made wrong turns. We will use that information to resolve what we want to do more and less of this coming year.

"Jackson and I will be facilitating this meeting. Each of you has an important role to play – to give us your honest beliefs in a respectful way. We also want your full undivided attention so put away your laptops and phones unless you have a real emergency. This could get a little rough and we need you fully engaged. Jackson, would you please go over our plan for afternoon?"

Jackson rose from his seat at a table near the front and walked over to the visual agenda on the wall (see Figure 19.1). He silently put a check mark in front of Introductions and Objectives and began to walk through the remaining items.

"We are going to start by arranging into groups for this afternoon. Pam and I have assigned each of you to a table. Please look on the back of your

MEETING TIP

Assign tables or free-for-all? It depends on your goals. In this nominal group exercise (in Appendix 2, page 269) the goal is diverse idea generation for the full team, so assigned.

Retrospective Agenda

✓ Introduction and Objectives
o Create Timeline (Groups)
o Dot Vote on Focus Events
o More of/Less of Based on Focus Events (Group)
o Resolutions for Our Team
o Communication Plan & Next Steps

FIGURE 19.1
Pacifica retrospective visual agenda.

name tag for a table number, and take a moment and move to your assigned table. We wanted to construct teams with diverse representation so the dialogue at your tables is more interesting."

Everyone looked at their name tags and rearranged themselves in generally compliant good humor. "Thanks, that was easy," Jackson patted the team metaphorically on the back. "Our first task is going to be to create a timeline of events for the program."

Herb took advantage of Jackson's pause to ask, "Do you want me to just do that, Jackson? I brought the final project plan with all the dates and tasks."

"Hold off for a moment, Herb, I'll get to instructions of the timeline creation in a moment. It's not just a timeline as you might think of it. OK?" Jackson got an OK from Herb so he continued.

"The timeline is going to have key decisions and milestones. We are then going to do a dot vote on which of the events were the best things we did, and which were the most problematic. We'll then have perhaps a dozen meaningful events to discuss in more depth at each table with a mission of extracting items to do more of/less of going forward.

"We'll take the more of/less of items and put them on the wall there," pointing to a more of/less of grid (see a grid example in Appendix 2, page 269). We'll do one more dot vote for most important more items and less items, talk that over, and make a resolution for our team.

"We'll wrap up the day by deciding how we want to communicate our results and then head out to the lawn for our celebration."

Pam stood and did her process understanding check. "Any questions or comments? Herb, does that answer your question? We don't want a sterile timeline. We want to create our team's meaningful event timeline."

Herb was on board, so Pam took the podium back over and introduced the timeline exercise.

"Here on the wall we have drawn out a timeline from a year ago July, when the studio was just an idea, to now. We'd like each table to come up with no more than 10 meaningful events or decisions that were important to this project. Once I say 'Go!' each of you will have 10 minutes to jot down your events on sticky notes. Then each team will select a spokesperson, and you will go around your tables one person at a time reading their cards and adjusting with the team at the table.

Once you are done sharing your cards with each other, you will combine or adjust as you like, and then pick the top 10 to share with the full group. You'll have a full hour to that. Then we will re-assemble and create our joint timeline.

"I have an example for how this should work. Jackson, could you bring over the cards you created earlier?" Pam began her tool advertisement (see Appendix 2, page 272). "First, I'll read a card I created. 'We decide to build a fuller plan' is the title. Under 'Description,' I wrote, 'Mary joined, and helped us decide that the simple scrum release plan was not enough.' Date is last January. To me this was a huge thing for the team, I don't see how we could be here today if we had stuck to the simpler approach."

Jackson took the cue as Pam ended. He took the first of his cards and said, "This card refers to that first meeting with Mary as well, where the team fully rejected my coaching." Jackson good-naturedly winked and smiled. "I would declare this a duplicate of the card Pam just read, combine it with Pam's card, and likely support the card as one of our top 10. Then I would go to my next card. This one says, 'Interviews refine product definition.' My description is, 'Team interviews 30 customers and prospects in six countries, learns what is really needed versus just guessing.' This is last year September through October."

"Good, Jackson," Pam ribbed her good buddy. "One last line to draw out for you. The end game is the items we want more or less of, or just to be sure we keep doing as we are. These two cards might wind up as resolutions to plan even more intensively, and to continue to engage with customers and prospects directly.

"Enough explanation?" Pam asked as she scanned the room for confusion. There were a few comments and clarifying questions and then the exercise began. Pam and Jackson each joined a team and began writing out their cards.

To allow Pam and Jackson to participate in the retrospective, Mary had volunteered to facilitate the remainder of this exercise, up to the Communication section. Mary had been an important influence on the project but has

MEETING TIP

It is very difficult to both participate in and facilitate a complex exercise like this. But it's also difficult to facilitate with no knowledge of the people or context. There's no perfect answer to this one.

purposely maintained some distance. Her role was not to do the work, it was to advise the team and consult with Sai and other leaders. Because the purpose today was to make resolutions for the next year, it was important the people responsible decide what to do for themselves. Mary felt certain that this team was ready to do that.

Mary walked around the room, checking on cards, giving occasional pointers on titling and generally supervising. She hadn't been with the team over the holidays, and since New Year's had only worked with Pam and Jackson to help with this retrospective. She felt proud of the growth of this team and excited for what they had accomplished. She was a little nervous about what the retro would conclude, as it was her first consulting gig.

In surveying the room Mary noticed that none of the executives were present (Sai, Heather, Melanie, Max). Jackson had felt strongly that the top brass shouldn't be invited to retrospective – that they weren't close enough to the work, even though they were obviously important to the project. The direct team members themselves needed time and space to retrospect – the agile self-managed team concept. This view had carried the day, so Gary and Sara were the most senior executives present. On the other end of the seniority spectrum, Jackson had advocated a broad sweep of team members, and Mary was on board with that as well. The purpose was to agree on a common view of project history and on improvement focus going forward, so the more of the team who could attend, the better. The large group demanded a clever meeting design which Pam and Jackson had delivered.

As the time allotted to generate the events at each table ended, Mary called the group to order and put them to work building out the timeline. Judy (User Experience Manager) was first to share her events. She stood in front of the timeline with a handful of stickies and began.

"'The interview in Bangalore that showed even a very small business could find this product compelling,'" she read. She posted this in last August and turned to the next card. "'The project structure meeting. It got us organized and communicating better. We needed that on top of the daily all team scrum.'" Judy posted this in February, and then continued through her 10 cards. Once she was done, each team representative in turn posted their items; more than half were duplicates so the total number of events on the board remained manageable.

Once the timeline of events was established, Mary asked for everyone to gather round the timeline and asked for a volunteer to do a walkthrough for the group based on the events on the board. Mary selected Vivian to do the job. Vivian started at the proposal to create the product, cruised through the D&G consulting team proposing agility to Sai and his boss and the connection to this project, through the creation of the studio, the team kickoff with Jackson, Mary's influence on the big planning project, ending with the CurrX disappointment and then the successful but small pilot.

Mary walked over to the visual agenda (see Figure 19.2) and checked off the Create Timeline item and explained the events to follow. First the dot vote to identify the most compelling events, followed by re-assembling in teams to identify the proposed more of/less of items. Then consolidating and sharing them, making some resolutions, and then agreeing on how to communicate the results. Mary then got out the dots for voting (Mary liked dot voting – it got everyone up and around and participating), handed them out, did a tool demo on how to vote, and then stepped back.

A few hours later the retrospective was done. Sai, Heather, Melanie, and Max had arrived together and listened intently to the resolutions the team had made for the year ahead. As the team broke up and headed to the celebration outside, Mary surveyed the plastered walls and the room in disarray and considered how the past few hours reflected on her work with the Pacifica team.

Mary was proud of the team and of herself. She hadn't been sure that she could be effective or fulfilled in an advising capacity. She found the growth of the team members and the team as a whole rewarding in a different but compelling way, versus her past direct project ownership.

Mary had worked hard in this retrospective to be a neutral facilitator so that the team could speak for itself and set its own path forward. Mary's role was ending along with the D&G engagement; Sai and Heather believed Pacifica could proceed on its own for now. As the

Retrospective Agenda

✓ Introduction and Objectives
✓ Create Timeline (Groups)
o Dot Vote on Focus Events
o More of/Less of Based on Focus Events (Group)
o Resolutions for Our Team
o Communication Plan & Next Steps

FIGURE 19.2
Pacifica retrospective agenda part two.

room emptied Mary surveyed the more of/less of ledger on the wall and what she saw confirmed for her, Sai and Heather's confidence that this team was ready to succeed on its own.

She started with the more of list. It included items derived from a debate on why the CurrX delay had occurred, such as earlier and deeper engagement with vendor partners, and environmental requests such as more privacy options for studio team members. Mary was happy to also see a commitment to more planning ahead and more focus on technical capabilities and risks, and a desire to broaden the use of A3s.

The less of list was similarly diverse, ranging from global (don't set firm dates before planning is done) to annoyance (no unannounced VIP visits to the studio).

The keep doing list brought Mary the most pride. At the top of the list was integrating events, to which the team gave the most credit for their success. There had been a 10-minute discussion just on these, with endorsement from developers, testers, and – perhaps most surprising – Sara. Sara noted that as product manager it had in the past been difficult for her to understand the technology and the details of her system, but the in-depth events made it easy for her to participate and learn. The events added a comprehensive closeness of the technology to the rest of the team.

Other items on the keep doing list were architecture simulations and team planning meetings, the workstream and governance structures (with a few minor tweaks), and the technical/marketing leadership partnership of Gary as chief engineer and Sara as product owner.

As Mary wrapped up her review and headed outside to join the team celebration she was intercepted by Jackson Maxim, the enthusiastic agile coach from D&G. Mary's relationship with Jackson had started out a bit rocky as she intruded on his agile coaching space and pushed him beyond the cookbook scrum process he knew, working to generalize the team's leadership approach to deliver rigor, alignment, and efficiency through frameworks. Fortunately Jackson had been open to learning and had been an excellent team member.

"Mary," he said, "this is my last week here. I'm heading off to another client. I wanted to just say thank you for helping this team succeed and for all you've taught me. Not just good agile techniques, which will make me a better coach. I'm going to do architecture simulations forever! More importantly you've taught me how to be a better leader."

"I'm sure the team will miss you, Jackson, you've helped them so much. Let's stay in touch," Mary smiled back at him. Even though Mary wasn't a hugger she couldn't

LEADERSHIP TIP

It's sometimes OK to hug.

resist. Hug over, Mary and Jackson ambled over to the hubbub on the lawn and got in line at the bar.

Signposts	• The Pacifica project has successfully released, in a highly limited way, MyHedge. They gather to do a full release retrospective. Led by Pam, Jackson, and Mary, they align on a common interpretation of the past and commit to how they want to work together more effectively in the future. • Jackson has now completed his agile coaching assignment and says goodbye to the now self-sufficient team and to Mary, who had become a mentor and teacher.
Leadership Guides	Do retrospectives more than just every sprint. Those can be valuable but the sweep of events is too short to catch the most important items. Spend some serious time periodically in extraordinarily well-prepared retrospective meetings.
Coming Up Next	Our engagement with Pacifica is over, our background and demonstrations of "People over Process" is complete. We wrap up with some final considerations.

Conclusion

People and interactions over process and tools. It's the very first agile value from the Manifesto and the most important. Yet, over and over, the focus in agile implementations is on process: scrum, SAFE, tools, sprints, user stories, and DevOps chains. Of course process is important – the Manifesto is clear on this. But the real leverage point, the key to sustainability, the driver of agility, is people and their interactions.

Why the focus on process? I would contend it is because people are messy; process is precise. Until now there has been no easily comprehensible model of how people and their interactions should change to support agility. Now we know.

- Agility is driven by people exercising facilitative leadership.
- Leadership is people helping teams be rigorous, aligned, and efficient.
- Leaders use frameworks. These include scrum tools plus architecture simulations, project plans, A3s, solution ranking, 2 × 2 matrices, and many more, including frameworks newly invented for specific situations.
- Organizational leaders have special obligations to help teams form, evolve, and connect.
- Extraordinarily well-prepared meetings are key fulcrums to efficiently gain rigorous alignment.

What concrete steps can you take to improve your own and your organization's leadership capabilities?

- Sponsor and organize extraordinarily well-prepared meetings.
- Be a demanding but respectful meeting participant. Insist (gently) on clarity on outcomes and paths.
- Demand and demonstrate rigor in all decisions. Options and facts.
- Learn to do A3s.

- Do a failure modes and effects analysis on an important system.
- Think "People over Process." When undertaking a new initiative don't over-focus on requirements or cost estimates; get the people and their interactions right first.
- Buy lots of copies of this book and distribute widely ;-).

Best wishes on your leadership journey.

Appendix 1

Manifesto for Agile Software Development

VALUES

We are uncovering better ways of developing software by doing it and helping others do it. Through this work we have come to value:

Individuals and interactions	over	processes and tools;
Working software	over	comprehensive documentation;
Customer collaboration	over	contract negotiation;
Responding to change	over	following a plan.

That is, while there is value in the items on the right, we value the items on the left more.

PRINCIPLES BEHIND THE AGILE MANIFESTO

We follow these principles:
- Our highest priority is to satisfy the customer through early and continuous delivery of valuable software.
- Welcome changing requirements, even late in development. Agile processes harness change for the customer's competitive advantage.
- Deliver working software frequently, from a couple of weeks to a couple of months, with a preference to the shorter timescale.
- Business people and developers must work together daily throughout the project.
- Build projects around motivated individuals.

- Give them the environment and support they need, and trust them to get the job done.
- The most efficient and effective method of conveying information to and within a development team is face-to-face conversation.
- Working software is the primary measure of progress.
- Agile processes promote sustainable development. The sponsors, developers, and users should be able to maintain a constant pace indefinitely.
- Continuous attention to technical excellence and good design enhances agility.
- Simplicity – the art of maximizing the amount of work not done – is essential.
- The best architectures, requirements, and designs emerge from self-organizing teams.
- At regular intervals, the team reflects on how to become more effective, then tunes and adjusts its behavior accordingly.

Appendix 2

Tool Tips

A3

Used in the Pacifica story: Pacifica Teleconference; Use of an A3, Chapter 17 page 225.

The label A3 comes from a size of paper, somewhat larger than 8 ½ by 14 inches, used in Japan. The A3 content was apparently routinely printed out on this large size because it allowed space to present a full problem statement and analysis on one page, complete with diagrams and charts. The label has grown to cover both the format of a set of documents, and the process of developing and socializing them. The best English documentation I've seen is from ex-Toyota executive John Shook. Durward Sobek, an engineering professor, has also written about and researched the practical use of the A3. There is plenty of information on A3s around, just Google. I can think of few improvement levers more powerful than teaching and encouraging the use of this tool.

The basic idea is simple. An A3 is a one-page problem-solving document and the cultural processes around its socialization. Its broad use builds rigor, alignment, and efficiency – a real boon to strengthening agility, at very low cost.

Let's start with the A3 itself. Figure A.1 is a sample A3 from a random website I found on the web through a Google image search. It will suffice.

The A3 is laid out in landscape mode in two columns. Major topics are confined to boxes, clearly labelled as to purpose, usually two to four boxes per column. The sample has three boxes in each row, laying out a common problem-solving sequence:

- Problem Statement.
- Target Statement (also commonly called Target State, Objective, or the like).

- Analysis (includes current state analysis, sometimes called Current State).
- Proposed Actions. The excellent facilitative leader will ensure that somewhere, whether in the analysis or the proposed actions, there is consideration of options using evidence and logic.
- Action Item (or Schedule, or Next Steps).
- Check and Act (or followup, checkpoint, etc.).

Screenshot of http://taskoconsulting.com/tag/kaizen-a3-sample/

In each box, since space is limited, the goal is to convey the information as clearly and in as few words as possible. If a picture can help, use a picture! The example has a few charts but typically there would be some diagrams or photos or other compelling visuals.

The A3 can be modified to be a short project initiation document as well as being used for problems and opportunities. I like to include a box on people, of course, identifying the team members and the governance level as well.

Often an A3 will grow beyond one page with more explanatory information, but the goal is to have additional material as "dig-deeper" but not intrinsic to the story being told.

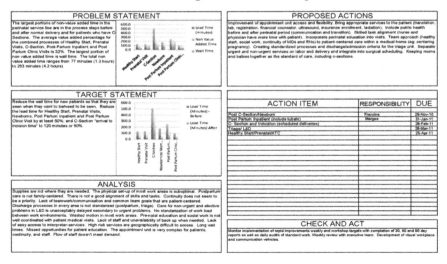

FIGURE A.1
A3 format.

The process of creating and socializing an A3 has a pleasant-sounding Japanese name: *Nemawashi*. An official site of Toyota UK (http://blog.toyota.co.uk/nemawashi-toyota-production-system as of October 15, 2018) defines *Nemawashi* as:

> **Nemawashi** (English: Laying the groundwork or foundation; building consensus): The first step in the decision-making process. It is the sharing of information about the decisions that will be made, in order to involve all employees in the process. During *Nemawashi*, a company seeks the opinion of employees about decisions.
>
> Literally translated as "going around the roots," particularly in the sense of digging around the roots of a tree to prepare it for transplant.
>
> Within the Toyota production system – and Japanese culture itself – the word has come to mean an informal process of laying the foundation and building a consensus of opinion before making formal changes to any particular process or project.
>
> Successful application of *Nemawashi* allows changes to be carried out with the consent of all parties.

The A3 format gives us rigor and some efficiency by forcing concise statements of problems and solutions.

Nemawashi gives us alignment and some efficiency by building consensus quickly and around clear statements of issues and facts.

In practice, *Nemawashi* is quite simple. As a problem is being understood or an initiative being formulated, a leader will begin to build out an A3. As soon as a testable problem statement is generated, that leader can share the draft, incomplete A3 to gain perspective, evidence, and alignment around the problem statement or opportunity. As the sections get completed, at whatever pace, the drafts can be shared and input and evidence and perspective and alignment gained. Sometimes drafts are shared at each section completion, and sometimes not until complete. When complete, then, the A3 will not only be a concise, clear, rigorous statement of a problem and a solution, but at least as importantly one around which the team is already quite aligned.

AGENDA

Described in Chapter 5 Background: Extraordinarily Well-Prepared and Conducted Meetings, page 43.

As boring as this sounds, the agenda is the one of the most important meeting tools. Think of it as the written detailed roadmap of the meeting, useful for the journey planner as well as for all the participants.

The format I'll show here is a useful standard format, but like all of the tools, I encourage leaders to experiment and customize for your own needs. I'm using the example of Mary's first meeting with the Pacifica team, from Chapter 6 Pacifica: The Course Correction Meeting. That was a long meeting with many participants and some well-

Sample Agenda: Pacifica Course Correction

July 23, 20XX

Pacifica Bank Headquarters, Tokyo Conference Room

Objective of Meeting: Review state of project and agree on way forward, as we turn towards developing the solution

Expected Outcome: Alignment on basic shape of project, specific next steps, and communication plan

Time	Activity	Facilitator	Outcome
9:00 – 10:00	Meeting setup: introductions, align on objective, outcomes, and meeting path	Mary O'Connor	Team understands and agrees on plan for the day and is ready to jump in and get to work
10:00-11:00	Understand Current State and Assumed Way Forward. Jackson will present current status and next steps according to the scrum agile methodology being followed.	Jackson	....
....	....	....	....

Attendee	Project Role	Primary Role in this Meeting
Damien Lopez	Test Manager	Begin drafting Master Test Plan; advocate for test driven development
David Phillips	Architect	Share state of architecture; advocate for appropriate steps and deliverables for effective tech arch
Gary Sunderland	Development Lead	Ensure proper flow of information from requirements to code, and associated people and organizational needs
....	....	....

FIGURE A.2
Sample agenda.

designed activities, so the agenda is quite detailed. Simpler and shorter meetings can have simpler agendas but don't skip.

The agenda header should show basic meeting information, such as date and location, participants (if not too many, else do a separate handout), and importantly, the meeting objective and desired outcome. In this case the roles of the meeting participants are important and not clear going in, so a section on roles is included. Then the timeline for the meeting is laid out, with a title for each activity, a description of each activity, and the expected outcome of each activity.

In some cases the meeting planner might do this kind of detailed agenda to help lay out the plan but share something smaller with the team, even just a visual agenda on the wall. The important thing is to have a great plan for the meeting and communicate it well enough to the participants so they know where they are going, why, and how they are going to get there.

CHECKING FOR ALIGNMENT

Fist to Five

First appears in Pacifica story, Chapter 6, page 60.

This is a quick and effective way to let meeting participants register agreement or concerns. Just ask participants to hold up their hands with five fingers for full agreement, fist for strong concerns. Then the facilitator or "fist to five requestor" can probe answers as seems appropriate.

Thermometer

In Pacifica project planning meeting, Chapter 10, page 139.

The first time I used this tool I used Alta Vista to look for a photo of a thermometer to make into a poster (i.e. a long time ago, before Google ...). I've used it many times since and like the visual impact.

You need to create a thermometer prop, on which you can dot vote. Here is an example of a thermometer created to get the team's perspective on a project plan. You can make any kind of visual, from a printout to drawing by hand on a chart or a whiteboard. You can also use whatever you like for the measures – here I've shown words, but you can also do percentages if that fits for you. (I don't like percentages much, because I always ask: "if it's 50%, what is the numerator, what is the denominator?" and get at best very complex or vague answers, or both.) You hand out dots, usually one per person, and let folks come up and register their opinion.

FIGURE A.3
Dot voting thermometer.

DOT VOTING

In Pacifica project planning meeting, Chapter 10, page 139.

This is one of my favorite little techniques. It gets people out of their chairs, fosters thinking and discussion, and gets measurable feedback.

In its simplest form, you have a display with choices, and each participant has some sticky dots. Everyone gets up and places dots on the display to indicate their view. You count up the dots for each choice and draw conclusions.

This is a common meeting technique and can be quite fun and effective. It can also be over- and poorly used. Some tips and variations to consider:

- Make sure you have a good question on which to vote. It has to be clear. One of my favorite dot votes is confidence in a plan: Are we 50% confident, 75%, or where are we? But what does that percentage mean? What is the numerator and the denominator? This is why, in the project planning meeting, Mary took special care to define the meaning of the percentages in words as well as number.
- Be clear in your mind, and with the group, what exactly the dot votes mean. Is it simply a vote and the largest number on a choice wins, like an election? Or is it a set of inputs meant to feed discussion around a consensus decision? Let's say we brainstormed on places to eat dinner and then do a dot vote. The winner is a Thai restaurant. But one of the voters is allergic to peanuts!
- Choose the number of dots and the number of permitted dots per selection. One dot per voter will reveal the top preference of each voter but not strength of commitment. Five dots per voter, one dot per selection will reveal the breadth of feeling but not much about depth. Five dots per voter, multiple votes per selection allowed, will allow depth of commitment to be displayed, which adds a nice dimension assuming that there is discussion to follow the votes and consensus development to follow.
- Manage the order of voting if needed. If your group has authority or sycophancy issues, it might be best to get the leaders, formal or informal, to go last.
- Use colors to distinguish voting groups.

EVALUATION MATRIX

The Evaluation Matrix is a common tool in comparing multiple options against multiple decision criteria. It's often seen in evaluating request for proposal responses against the stated requirements. Let's use that for our example, shown in Figure A.4.

In this case we are a small team preparing for our two-day architecture simulation and project planning meeting. The rows list out the evaluation criteria (close to office, quality of room), while the options are in the columns (our cafeteria, downtown hotel). The cross-hairs at the intersections contain our evaluation of the degree to which the option meets the requirement. Depending on your needs, you can evaluate as none/some/fully. Use a numeric scale which gives the added benefit and risks of enabling quantified comparisons, or whatever you like. I've even seen icons of animals, smiling or frowning suns, you name it!

In Figure A.4 we have a couple of special columns:

Criteria Weight

This is useful to spark discussion (and alignment) around which criteria are most important. If you are doing numeric evaluations you can use the weights to calculate a summary result. If not you use this just to drive discussion and alignment. I'd suggest that you keep your weighting simple – I like something like high, medium, low, or 3, 2, 1. I've seen this get very complex for an RFP response for a new system where

	Weight	Our cafeteria		Municipal Convention Center		Downtown Hotel	
		Raw Score	Weighted	Raw Score	Weighted	Raw Score	Weighted
Close to Office	2	3	6	3	6	2	4
Room Quality	3	1	3	2	6	3	9
Food and Beverage	1	1	1	2	2	3	3
Cost	2	3	6	2	4	1	2
Network	3	3	9	1	3	2	6
Raw Total			25		25		24
Indexed Total			56		56		53

Weights: 3= Very Important, 2= Important, 1= Nice to Have
Scores: 3= Best, 2=Middle, 1=Least Attractive

FIGURE A.4
Evaluation matrix.

a hundred or more criteria are used and the weights range from 1 to 10 and even decimals are allowed!

Summary Totals

The totaling is done by multiplying the weight times the score. I like to do an indexed percentage where you take the totals for each solution and divide them by total points possible. That is what is shown in the example.

Numeric ratings provide the added benefit and risks of enabling quantified comparisons. I say "and risks" because the value of doing this exercise is to decompose, analyze, and discuss; rarely is a complex comparison of options a simple mathematical exercise. Whatever criteria you choose, and whatever weighting and evaluation scales you use, the totals cannot adequately reflect real-life values, weights by decision-maker(s), and gut feel which remains important. The risks are thus that some participants may see the group consensus ratings as the "team recommendation" and feel betrayed if management over-rides or adjusts the results.

Numeric ratings at the cross-hairs can be simple (3 = fully meets, 2 = meets, 1 = gaps, 0 = fails) or complicated (10 = exceeds without customization, 8 = meets without customization, 5 = meets with customization, 3 = meets with committed development, 1 = on road-map but not scheduled or committed, 0 = does not meet, with decimal points allowed). You can also stack-rank the options as in the example shown.

FAILURE MODE AND EFFECTS ANALYSIS

In Pacifica story, Chapter 14, page 202.

Failure modes and effects analysis (FMEA) is a common tool to identify potential failure points, and by identifying prevent, detect, and be prepared to repair failures. FMEA is a powerful and simple tool of which many agile teams may be unaware. FMEA is taught in most engineering programs but not applied consistently enough to our system designs. (There is a nice summary of FMEA in Nancy R. Tague's *The*

Quality Toolbox, Second Edition, ASQ Quality Press, 2004, pages 236–240.)

The FMEA tool is simply a table, either in Excel or Word or the like, that begins with a list of each thing that can go wrong (Failure Modes). These can be attributed with probability and severity of failure to help focus limited resources on the most important failure modes. Other columns can include the effects of the failure, local and system-wide; how the failure would be detected; countermeasures to prevent failure; repair steps; and time to repair.

Let's look at a simple example. Figure A.5 shows a row from an FMEA focusing on the database. The database stops or slows down for several reasons: a bad release, human error, or hardware failure. The local effect is failed queries, while the system effect is catastrophic. Severity is rated as very high and the probability absent controls moderate; these ratings will certainly draw focus to ameliorating this failure point. Controls are identified to include testing, the DBA process, and the existence of passive node for failover. Detection would be immediate from real-time monitoring or slower via help desk reports. Finally, recovery from a failure could be done via backing out a release of bad code or failing over to the passive node.

The columns can be adjusted to fit circumstances. Much more information on failure mode and effects analysis is widely available in engineering literature.

Introducing the FMEA and helping a project team through its use is an excellent example of facilitative leadership. This framework efficiently promotes rigor and alignment.

FIVE WHYS

Five Whys is a popular lean technique aimed at getting to the root cause of a problem so that it can be definitively repaired. It was reportedly developed at Toyota and is taught as part of Toyota's problem-solving training. Five Whys is less a standard technique than a reminder to dig deeper than the surface problem. A good example of this in the book is the deep dive Lois helped (or forced) her team to do in the Saga of the Checkboxes (Chapter 18, page 238).

Component	Failure Mode	Causes	Local Effect	System Effect	Severity	Probability Absent Controls	Controls	Detection	Repair
Primary Database	Stoppage or Slowdown	Bad release; human error; Hardware failure	Database queries fail; no data provided to app	Dead in water	Very High	Moderate	Release mgt (testing, etc); DBA processes; passive node failover	Real time monitoring; help desk reports	Backout; fix in production; failover

FIGURE A.5
Failure mode effects analysis chart.

The classic example of Five Whys involves first asking why a problem occurred. Then, keep asking "why?" until there is no more to be learned and the underlying solution is divined. There is no magic in the number five – it's just a handy heuristic aimed at encouraging deeper inquiry and understanding. Once a team is trained in Five Whys and has seen it used a few times, there is a tendency to keep digging. I've seen teams in problem meetings look up at each other, say "Five Whys" and continue questioning.

An example from Taichi Ono, considered the father of the Toyota production system aka lean manufacturing, is included in his book *Toyota Production System* (Productivity Press, 1988, New York).

The machine stopped working

1. Why did the machine stop?
   a. There was an overload and the fuse blew.
2. Why was there an overload?
   a. The bearing was not sufficiently lubricated.
3. Why was it not sufficiently lubricated?
   a. The lubrication pump was not pumping sufficiently.

4. Why was it not pumping sufficiently?

a. The shaft of the pump was worn and rattling.

5. Why was the shaft worn out?

a. There was no strainer attached and metal scraps got in.

Without diving through the "Whys" workers would simply replace the fuse or the pump and the failure would recur. Instead, the pump was replaced with a strainer, and the machine would be more reliable.

This example is straightforward and doesn't get into organizational and cultural issues. In large-scale technology, it's common that by the second or third "why" you'll be into areas that may not be able to be influenced. At that point leadership might consist of doing what you reasonably can.

MORE OF/LESS OF

In Pacifica retrospective Chapter 17, page 245.

This is an easy, highly visual exercise that can be combined with other techniques such as nominal group and dot voting. It is useful in retrospectives.

Create a visual display that looks like this:

More of	Less Of

FIGURE A.6
More of/less of wall.

Generate cards/stickies to put up in the columns. You can do this in small groups with conversations, solicit as a facilitator (this violates the "make them do the work" principle but it's fast and easy), have individuals each do five of each or whatever seems like it will work with your group. If you have a lot of stickies in each column, assign a few attendees the task of sorting into categories/de-duplicating. If you have the participants do the cards, have the writers explain the idea and why it's valuable. You can plan next steps from there, do dot votes on the top ones to try to implement next, or just talk it over and see what team has energy around.

NOMINAL GROUP

In Pacifica project planning page Chapter 10 Pacifica: The Project Planning Meeting, page 115 and in retrospective page 245.

Kind of a funny name for a common technique. It gets overused and badly facilitated more than any other technique I see other than the "Facilitator at flip chart writing stuff down" non-technique. I don't know how many bad meetings I've been in where we are split into groups to ostensibly do something and then report out that resulted in nothing at all – I don't even want to think about it. Doing this badly definitely violates our efficiency principle, "don't waste our time!"

That said, it remains a valuable tool. The example of the project planning exercise at Pacifica is an example of a good use, where Mary breaks the larger group up by functional area and has each subgroup lay out their project tasks for integration back into the larger group. Another good example is in the Pacifica retrospective, where groups were cross-functional and each focused on the same challenge. A bad use might be to break up a large group randomly, ask for ideas on how to lower costs for the company, get a list, prioritize, and think you've got something – probably the wrong people, no evidence, really just a brainstorming exercise dressed up to look like something more.

There is plenty of documentation on Nominal Group Technique (NGT) available. You can start with the Wikipedia article which is a good overview.

NGT can be useful for a group to generate ideas and organize them, especially for large groups. A simple explanation:

* The larger group (say, 20 or more people) is split into smaller groups for discussion and idea generation. You can do this randomly, by who is sitting at which table, by sales region, by function, by department, whichever way will get you the results you are seeking most effectively.

- The facilitator explains the challenge, the process, and does a tool demo.
- A quiet period is imposed so each person can think and prepare some cards.
- Each subgroup has a facilitator/reporter identified.
- The subgroup goes around the table, reading one card at a time, clarifying meaning, de-duplicating in the group before reporting out.
- When all the groups are done, the groups report out one at a time. The disposition of the report out depends entirely on what the goal of using the technique is. See the project planning chapters (page 115) for one good example.

Some characteristics of the exercise include:

- The quiet time encourages everyone to provide input – alignment.
- The de-duplication at the table reduces waste – efficiency.
- If set up well, it can help to generate options and arguments – rigor.

NGT can be valuable but don't waste it.

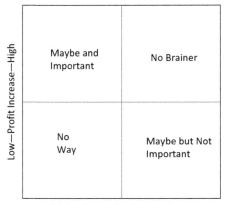

FIGURE A.7
Two by two quadrants.

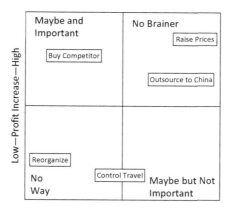

FIGURE A.8
Two by two ranked options.

TOOL ADVERTISEMENT

Pacifica's use of tool advertisement page 127.

A tool advertisement is an amazingly effective technique that helps exercises in meetings succeed. Before beginning an exercise, the facilitator simply does an advertisement for the tool – explaining the tool and giving a demonstration of how it will be used.

Imagine I'm facilitating a retrospective using a more of/less of exercise. I'll begin by setting up the wall with masked tape and signs, or writing on the whiteboard, as shown in the more of/less of exercise. Then I'll explain the exercise. Now, here comes the core of the advertisement: I'll demonstrate using a sample. For example:

Lets say we are doing this after working in my yard for a few hours. We might write "more of" cards of "take regular breaks with lemonade" or "wearing gloves" or "be sure it's a nice day." We might do "Less of" cards of "playing Meatloaf music" or "getting sunburn because I didn't wear a hat."

I would write up cards either in advance or real time and post them visibly in the columns. Then I'd review them, let folks stare for a moment, and then check to be sure everyone understood what we were going to do/take any questions. Then, off we go!

There is a good tool advertisement example in the Pacifica project retrospective section starting on page 247.

This is one of the tips folks who go through training forget or do not do the most.

TWO BY TWO

The two by two is a classic analytic technique that provides a simple and compelling model for rigorous and efficient alignment. The idea is to take two concepts and create a matrix. A sample is provided in Figure A.7, where the two concepts are cost to implement and profit created. The Gardner Group's Magic Quadrants two by two similarly cross-hatch ability to execute and completeness of vision to evaluate competitors in a market.

The example given is an excellent way to get a group to rigorously evaluate and rank ideas for improvement – of anything, really – but, in the realm of software development, of most use for improvements to project processes during retrospectives, or system performance and stability, or cycle time of application to close. The team could brainstorm ideas, or develop them via nominal group, and then place in the grid. You can simply place within a grid, or use the scales to more fine tune gradations as Gardner does in the Magic Quadrants. Figure A.8 shows the prior two by two grid populated using gradations.

I've seen quite a variety of axes in two by twos. Some include time to implement, numbers of departments involved, risk, completeness of solution, benefit to operations, benefit to customers. Anytime you need to rank a list of options in a simple and compelling way, the two by two is for you. If you need to rank on more complex scales, try the evaluation matrix.

Index

2 x 2 matrices, 218, 272–273

A

A3 process, 218, 225–230, 257–259
adaptive process control, 8–10, 19
agendas, 260–261
agile
    emotional roots, 7–12
    facilitative leadership, 10–12
    organizational leadership, 13–14
    Pacifica case study, 3–5
    vs. planful management, 27–29
    scrum release planning, 103
    team composition, 146–148
agile management tools
    A3 process, 163–164, 218, 225–230,
        257–259
    alignment, 261–265
    dot voting, 139–141, 246, 249, 262–263
    efficiency, 31–32
    evaluation matrix, 264–265
    failure mode and effects analysis,
        202–204, 265–266
    fist to five, 261–262
    five whys, 266–268
    more of/less of, 246, 249–250, 268
    nominal group technique, 269–270
    thermometer process, 261–262
    tool advertisement, 272
    two by two, 218, 272–273
Agile Manifesto
    origins, 7–8, 10
    principles, 255–256
    team composition, 145, 146, 150
    values, 255
alignment
    decision on whether to call a
        meeting, 201
    facilitative leadership, 11–12
    meetings, 46–47, 48–49
    organizational leadership, 24–26

teleconferences, 224
    tools for checking, 261–265
architecture
    for agility, 71–78
    project planning meetings, 109–110
architecture simulation meetings,
        72–78
    conducting, 82–86
    Pacifica case study, 79–100
    preparing for, 73–75, 79–82
    reviewing the going-in architecture,
        86–91
    running scenarios, 91–97
    transition to next steps, 78, 97–100

B

belonging, 34

C

Chief Engineers, 19–20
code migration, 110, 111–112
consensus, decision-making, 25–26
course correction meetings, 51–67

D

daily scrum meetings, 195–199
decision-making
    consensus and dissent, 25–26
    organizational context, 24–25
    rigor, 23–24
demos, 205–213
developer-centricity, 7–8
development, Microsoft Solutions
        Framework, 154–155
development team member role, 159–160
dissent, decision-making, 25–26
documentation
    architecture for agility, 71–72
    as connector, 162
    demos, 206–207

Printed in the United States
by Baker & Taylor Publisher Services